The World of
VENOMOUS ANIMALS

Title page: A giant hairy scorpion, *Hadrurus* sp., a genus from the southwestern U.S. and northern Mexico.

To my grandsons: Danny, Michael, and Victor—*M.F.*

ISBN 0-87666-567-9

Distributed in the UNITED STATES by T.F.H. Publications, Inc., 211 West Sylvania Avenue, Neptune City, NJ 07753; in CANADA by H & L Pet Supplies Inc., 27 Kingston Crescent, Kitchener, Ontario N2B 2T6; Rolf C. Hagen Ltd., 3225 Sartelon Street, Montreal 382 Quebec; in ENGLAND by T.F.H. Publications Limited, 4 Kier Park, Ascot, Berkshire SL5 7DS; in AUSTRALIA AND THE SOUTH PACIFIC by T.F.H. (Australia) Pty. Ltd., Box 149, Brookvale 2100 N.S.W., Australia; in NEW ZEALAND by Ross Haines & Son, Ltd., 18 Monmouth Street, Grey Lynn, Auckland 2 New Zealand; in SINGAPORE AND MALAYSIA by MPH Distributors Pte., 71-77 Stamford Road, Singapore 0617; in the PHILIPPINES by Bio-Research, 5 Lippay Street, San Lorenzo Village, Makati, Rizal; in SOUTH AFRICA by Multipet Pty. Ltd., 30 Turners Avenue, Durban 4001. Published by T.F.H. Publications Inc., Ltd., the British Crown Colony of Hong Kong.

The World of
VENOMOUS ANIMALS

Dr. Marcos Freiberg and Jerry G. Walls

Acknowledgments

We are grateful to the following friends and co-workers for their assistance in parts of this book: Dr. Richard Alphonse Hoge, Director of the Herpetology Department, Butantan Institute, Sao Paulo, Brazil, and his collaborators, Dr. Sylvina Alma Romano and Dr. Pedro Antonio Federsoni; Dr. Persio di Blasio, head of the Poisons Department at Butantan Institute; Mr. Juan José Blengini, Instituto de Investigación de las Ciencias Biologicas, Montevideo; Mr. Rodolfo A. Lores Zaez, Montevideo Zoo; the late Dr. Jorge W. Abalos, Director of the Córdoba Serpentarium, Argentina; and Mr. Addalberto Ibarra Grasso, Buenos Aires. Mrs. Inés Pardal, Buenos Aires, translated an early draft of the manuscript from Spanish to English. Technical assistance for some of the photos was given by Mr. Rogelio Gutiérrez and Alcide Perucca, Buenos Aires. Special thanks are also due to Dr. Sherman A. Minton, Indiana University Medical School, and John H. Trestrail, III, Western Michigan Poison Center.

Contents

Introduction

Man's imagination has always been fascinated by the almost mystical properties of venoms that can cause pain and even death from the sting or bite of the smallest and most inconsequential animal such as a spider or bee. The most primitive tribes recognize that certain animals are not to be touched unless one is willing to take the risk of being bitten or stung. Unfortunately, overactive imaginations have transferred the danger from truly venomous animals to any similar type of animal, resulting in the dread of any snake or any spider by many people.

Throughout their long evolution, many different types of animals have developed different methods of securing food and fighting off predators. Venoms have been developed in many unrelated animals along with often complicated methods of delivering that venom into the prey or predator. Except for birds and crustaceans, there are virtually no major groups of animals that lack venomous species.

So many animals are at least theoretically dangerous to man in some way that it was necessary to restrict the coverage in this book to only those animals that actually produce a venom that can be injected to produce symptoms in a human. We will have to ignore those animals that are dangerous because their size and teeth allow them to take bites out of humans, as do sharks and many large mammals. Also ignored are the numerous animals, mostly fishes and molluscs, that cause serious poisoning when they are eaten under certain conditions. Even among the venomous insects and snakes there is so much diversity that we will only skim the surface and mention just some of the more familar or spectacular pests.

This is a book about animals; it is not a book about medicine. First aid methods and medical information are mentioned only casually for some groups and not at all for others. Many traditional methods of first aid for envenomation are currently under attack, some supposely tried and true techniques are now thought to do more harm than good, and there are major legal problems in suggesting treatments. In any case, the safest procedure after suffering a sting or bite that seems to be causing unusual symptoms is to get prompt treatment. Check the site of the sting or bite to make sure that no stinger or fang is embedded in the skin; if so, carefully remove it. If possible to do so safely, collect the animal that did the stinging or biting for positive identification; try to damage the specimen as little as possible. As rapidly and calmly as possibly, get the victim to the nearest source of medical attention. If you are in an area where certain types of snakebites or scorpionstings or beestings are prevalent, then there will probably be antivenin available at local hospitals. In most cases immediate application of soothing liquids such as alcohol, a mixture of ammonia and alcohol, or calamine lotion will do no harm. Additionally, most drug stores and supermarkets sell applicators of chemicals especially designed to calm the pain and swelling of common insect bites. Other than these basic steps, first aid is best left to the experts.

One last thing to be considered is the fact that what is a mild sting to one individual may be a serious problem to another person. Individual allergic reactions to beestings are common and well known, and it is probable that unpredictable allergic reactions explain the different symptoms that may result from stings or bites by the same species of fish or insect, for instance. This possibility should be considered whenever serious or unusual symptoms result from the sting or bite of a normally only mildly venomous or even nonvenomous animal.

In no part of the world are venomous animals so prevalent that it is not possible to safely enjoy the outdoors. Eventually, however, everyone will have an encounter with an at least mildly venomous insect, snake, fish, or marine invertebrate. There is no reason to become paranoid about venomous animals lying under every rock and lurking in every bush, but it can certainly do no harm to know what is out there.

1: Jellyfishes and Allies

Even the most primitive of animals long ago developed methods of protecting themselves from enemies and of efficiently securing food. Often this has included the evolution of venoms and methods of injecting venoms into potential prey or predators. The first many-celled animals to develop distinct tissues were the Coelenterata, the phylum including such animals as the familiar freshwater hydras, sea anemones, jellyfishes, and the corals. These animals have an outer layer of cells called the ectoderm and an inner layer, the entoderm, separated by a noncellular gelatinous or fluid layer, the mesoglea. They have no brain as such, although there is a complicated system of nerves that causes them to react immediately to touch, light, chemicals, etc. There is no anus, so the solid waste products of digestion are expelled through the mouth, the only opening into the digestive tract. As a whole, the phylum can be recognized by two characters: the presence of what is termed radial symmetry (a cut bisecting the mouth area in any plane results in two equal sections with an equal number of tentacles and other organs in each) and the universal presence of stinging organelles called nematocysts housed in special cells called cnidoblasts.

The nematocysts are the important factors in the venomous nature of several members of this phylum. Each nematocyst contains a coiled filament that is usually barbed and connected to sensory bristles that, when triggered, cause a lid on the cell to open, thus changing the pressure inside the cell. The change of pressure causes the filament to be everted. There are many types of nematocysts, but most of them contain a minute amount of venom. Because each tentacle of a coelenterate contains thousands or even millions of nematocysts, the danger lies in the number of stings received rather than in the results of a single sting. Although the stings of most coelenterates such as corals and

The tentacles of the Portuguese man o'war provide protection for several animals that have adapted to avoid their sting or that have developed immunity. The most familiar of these is the small fish *Nomeus gronovi*. Photo by Charles Arneson.

Facing page: Many hydroids can sting to some degree, but the small size of most species makes them easy to ignore. The hydroid *Lytocarpus philippinus,* however, can be several feet tall and may occur in colonies covering many square feet. This hydroid and several similar types can cause severe rashes and pain but seldom inflict serious injury on divers. Photo by Allan Power.

sea anemones are at most local irritants that may cause temporary discomfort, redness, and itching, there are a few coelenterates that can cause severe symptoms in humans and may even cause death.

THE PORTUGUESE MAN O'WAR

The Portuguese man o'war, *Physalia*, is a colonial coelenterate known as a siphonophore. Although at first glance it may look like a jellyfish, it is actually a pelagic colony of many different types of semi-independent cells each specialized to perform certain functions. Some cells only digest food, others only serve to inflate the blue float with nitrogen, and still others have the functions of protecting the colony and acquiring food. The transparent blue float or umbrella is often seen bobbing on the waves or stranded on the beach after storms and results in the common name of bluebottle for *Physalia*. The float is harmless—the danger lies in the numerous tentacles that may extend for 50 or more feet beyond the 8-inch float and contain millions of nematocysts. The tentacles are delicate and easily broken, but they stick to the skin even after being

Even though it is stranded and drying on the beach, a *Physalia* is still to be avoided. The nematocysts may be active for days after the rest of the colony is dead and decomposing. Photo by Keith Gillett from *Australian Seashores in Colour*.

In close-up the tentacles can be seen hanging from the edge of the float or sail in various degrees of contraction. Each "ring" on the tentacles may contain hundreds of nematocysts.

broken, and the nematocysts may live for several hours, continuing to inject poison into the predator—or human arm or foot as the situation may be. The stings are intensely painful immediately. As the pain and effects of the venom move up the limb from the actual stinging site, other complications arise, including respiratory distress causing difficulty in breathing and speaking. Victims with heart problems or other debilitating conditions have been known to die from massive stings. Normally, however, recovery is rapid, though painful.

If you see a Portuguese man o'war, leave it alone. Even specimens that have been on the beach and drying in the sun for hours or days may still have a few living nematocysts present in the tentacles. A dead bluebottle can sting almost as effectively as a live one. The usual remedies for stings, such as papain (as from meat tenderizers) or a mixture of alcohol and ammonia, may be tried to relieve the pain and neutralize the venom on the skin. Meat tenderizers work quickly and effectively if applied to the sting immediately, moistened slightly, and then rubbed into the welt; it should then be rinsed off as soon as the pain and any swelling subside. If any complications arise, visit the doctor immediately. If any pieces of tentacles adhere to the skin, remove them with sand or a gloved hand to prevent continuous stinging. Portuguese men o'war are dangerous animals and should never be taken lightly. Although the animal is common over all the tropical seas of the earth, it is typical of oceanic waters of higher salinity, such as those of southern Florida and Australia, where most stinging incidents occur. It has been found well north on the Atlantic Seaboard of the U.S., however.

As an aside, it might be mentioned that *Physalia* has its own problems. It is the preferred food of many small animals, including very specialized nudibranchs. The pelagic nudibranchs *Glaucus* and *Glaucilla* not only feed on the tentacles of Portuguese men o'war, but they are able to extract the uneverted nematocysts intact and transfer them into the various cirri and filaments on their own bodies. Here the nematocysts survive for long periods of time and give the nudibranchs the same potent stinging ability as they gave the bluebottle. Because of the small size of the nudibranchs, however, the most they can do to a human is cause minor pain and irritation.

Although most familiar anemones have at best a minor stinging ability, some are quite strong stingers. This bright orange *Anthothoe* from the Pacific can inflict a painful sting but is too small to be of much medical importance. Photo by Keith Gillett from *Australian Seashores in Colour.*

This rather amorphous growth is one variety of stinging coral, *Millepora*. These animals are actually hydroids that secrete a calcareous structure around the colony, much as do the true corals. The various species of *Millepora* are variable and occur almost worldwide in tropical seas. Photo of *Millepora squarrosa* by Dr. Patrick L. Colin.

Not all *Millepora* colonies are shapeless masses. Sometimes they are just encrustations on dead gorgonians or corals, and sometimes they are well-shaped branched colonies, as shown here. Stinging corals are considered great hazards by divers and can cause extremely painful and persistent stings. Photo by Allan Power.

Above: Two views of *Chrysaora melanaster,* a sea nettle jellyfish from California. Similar species occur in all oceans. Photo at left by Ken Lucas at Steinhart Aquarium; that at right by Dan Gotshall. **Below:** Two views of the lion's mane jellyfish, *Cyanea capillata,* a dangerous species that reportedly can reach a disc diameter of 6 feet. Photo at left by Robert Abrams; that at right by Dr. Leon P. Zann.

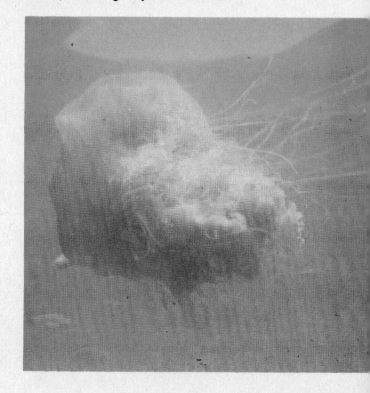

Chiropsalmus species (probably *C. quadrumanus*), a cube jelly found along the southern U.S. coast and in the Caribbean. Photo by Dr. Patrick L. Colin.

Two views of discharged nematocysts in a slide preparation. The large and numerous barbs are plainly visible. Photos by Dr. T. E. Thompson.

JELLYFISHES

Everyone recognizes a jellyfish. They look like bluish, rosy, violet, or transparent opened umbrellas with four, eight, or more tentacles dangling from the rim. The tentacles may be long or short, thick or thin, evenly distributed or in clusters. There are numerous species of jellyfishes found in all the seas of the world. Rocked by the gentle motions of the waves, they float on or near the surface of the water and often go unnoticed by man until there is a rash of stingings among bathers in an area.

Technically, true jellyfishes are just the sexual half of a rather complicated life cycle involving a sedentary, small, non-sexual stage that produces a free-swimming sexual stage. The non-sexual stage is called a strobilus and looks something like a small anemone or hydra. It buds off small jellyfish in an upside-down position, and as each matures it breaks free and floats off. The mature jellyfish has well-developed sex organs under the umbrella between the rays of the body; a fertilized egg from the parent jellyfish gives rise to a larva that settles to the bottom and develops into the strobilus. Both true jellyfishes and the free-swimming stages of certain marine hydra-like coelenterates are called medusae because of their fancied resemblance to the snake-locked Medusa of Greek mythology.

As in other coelenterates, the tentacles contain numerous nematocysts. In most cases their sting is relatively harmless, causing just minor irritation unless the animals are present in large numbers, as often happens. The common sea nettles, *Chrysaora* and *Dactylometra*, often invade swimming beaches around the world in vast numbers, inflicting painful and sometimes serious stings. Both *Chrysaora* and *Cyanea*, another common pest, can be found in relatively cool waters as well as the tropics, and incidents of jellyfish invasions are often noted in the newspapers. Some beaches even erect jellyfish nets around public swimming areas in an attempt to prevent the problem.

15

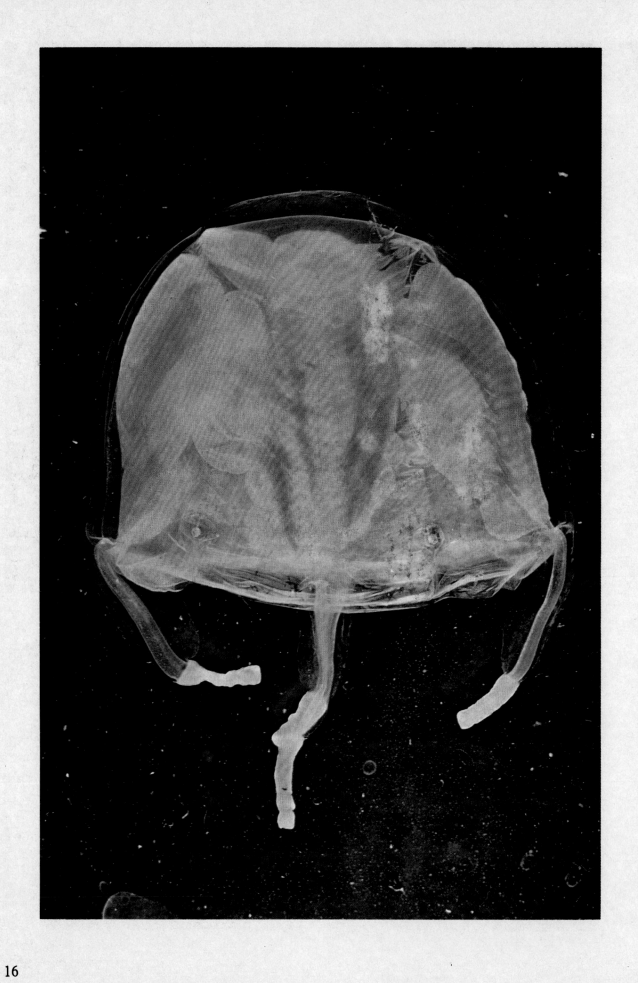

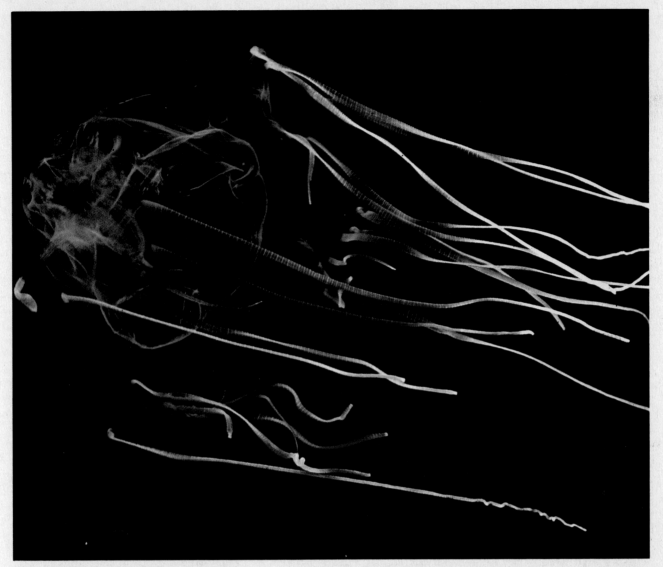

Swimming specimen of the very dangerous sea wasp *Chironex fleckeri* from Australia. This species has caused human fatalities. Photo by Keith Gillett from *The Australian Great Barrier Reef*.

Facing page: A contracted specimen of the sea wasp *Carybdea* sp. from the Bahamas. Photo by Dr. Ron Larson.

The tropical jellyfishes known as sea wasps or cube jellies are the real villains of the group, however. They are rather box-like in shape, often quite small, and have the tentacles in clusters at the corners of the body. Most species are tropical, but some occasionally are found as far north as the central Atlantic Seaboard in the U.S. The most common genera include *Chiropsalmus*, *Chironex*, and *Carybdea*. Their venom varies from mildly dangerous to being capable of inflicting death to a human (usually a child or elderly adult) within five minutes, so they are not to be ignored. The symptoms are much like those of bluebottle stings, and it is unlikely that healthy adults would die from casual contacts with the cube jellies. In Australia, sea wasps are considered a serious problem at some times of the year, when they invade public beaches in tremendous numbers. Large swarms have also been noted in the Caribbean, though the Atlantic Ocean species seem to be less likely to cause serious injury than the Australian species.

2: Sea Urchins and Starfishes

The serrated, sharply pointed spines of *Diadema antillarum* are familiar sights to divers in the Caribbean and are avoided as much as possible. The spines become more uniformly dark with growth. Photo by Dr. Patrick Colin.

Starfishes, sea urchins, sand dollars, sea cucumbers, and their allies are placed in a phylum called Echinodermata. The name of the phylum means "spiny skin," in reference to the hard calcium plates and rods embedded in the skin of most species. Like the coelenterates (to which they are not closely related), echinoderms have radial symmetry, though here the body is composed of five (or a multiple of five, rarely six or eight) arms around a central disc. The mouth is usually on the ventral surface of the body; there is an anus. Echinoderms are characterized in addition by a complex system of vessels and valves called a water vascular system. By controlling the water pressure within the body, it can operate tube feet used for movement and obtaining food.

Several groups of echinoderms have species with venomous spines, but the only ones we will discuss are two types of sea urchins and a starfish; the others are all rather obscure animals, and even the three types considered below are not widely known problem animals. Starfishes have five or more free arms that are attached broadly to the central disc. Sea urchins have fused the arms into the central disc, resulting in a more or less spherical or flattened body without free arms.

SEA URCHINS

Most sea urchins are extremely spiny, often with several types of spines on different parts of the body, the different types having various functions. Spines near the mouth may be used for locomotion, swiveling on ball and socket joints; others may function in excavating shallow burrows for the urchin. Spines on the dorsal surface, however, are usually for defense.

In the long-spined *Diadema* species found in tropical and subtropical oceans, the spines may be many times longer than the height of the body and are often segmented. They

Facing page: This brightly colored diademid sea urchin should be treated with caution even if it were not truly venomous, as the breaking off of the spines in the hands or feet would be painful even if venom were not present.

18

are extremely sharp and brittle. If stepped on, they can easily go through the sole of a shoe, the spines penetrating deeply into the foot of the unwary wader. Once in the muscle they break into many pieces, increasing the pain. It is still uncertain if venom is associated with the spines, but the pain they cause is intense whether venom is present or not. Secondary infections are common, and the wound is usually very slow to heal, as the pieces of spine must gradually dissolve in the tissue (usually producing a blue stain). Swelling and an aching pain may be present for several hours or even days.

It is not certain if *Diadema* is actually venomous or not, but there is no doubt that sea urchins of the genus *Toxopneustes* are dangerously venomous. *Toxopneustes* is a rather normal-looking sea urchin found in tropical Indo-Pacific waters. Not only does it have the usual rather heavy spines on the dorsum, but also the pedicellariae are very large. Pedicellariae are specialized spines used to remove debris from the surface of the body in urchins and even occasionally to capture small prey animals or parasites. In *Toxopneustes* the pedicellariae end in three wide, movable

An adult *Diadema setosum*, a venomous Indo-Pacific urchin. Photo by Allan Power.

Toxopneustes pileolus, a venomous sea urchin with flower-like pedicellariae. Photo by Keith Gillett from *Australian Seashores in Colour*.

Close-up of *Toxopneustes pileolus* showing the three-valved, red-centered pedicellariae among the blunt spines, the tube feet, and the normal pedicellariae. Photo by Walter Deas.

valves that are sharply pointed and toothed at the tip. Each valve has an associated venom gland and duct. When a cluster of sensitive bristles on the valves is touched, the valves snap shut. The teeth at the end of the valves penetrate the skin and the venom ducts start delivering their venom. The "bite" of the valves is so strong that they can actually be pulled from the body of the sea urchin when the hand or foot of the careless diver is pulled back in pain. Under these circumstances they remain shut and continue to deliver venom for up to several hours. Thus, the pedicellariae must be physically removed from the bite area to stop the envenomation. The venom causes intense pain at and near the site of the bite and temporary loss of movement of the arm or leg. Usually the pain subsides in a few minutes, but it may take hours for all symptoms to disappear. Death could result at least theoretically from drowning caused by the temporary paralysis, but usually the symptoms subside quickly. Remember that any adherent pedicellariae must be removed, however.

Acanthaster plancii, the crown of thorns starfish, is the only species of starfish known to be definitely venomous to humans. Other species may cause minor effects after handling, but this may be due to simple pricking by the spines. As can be seen in these photos taken in various parts of the Indo-Pacific, the species varies considerably in color and spininess. Photo above by Keith Gillett from *The Australian Great Barrier Reef;* photos below by Rodney Jonklaas.

A smaller version of the crown of thorns, *Acanthaster ellisii,* is found in the East Pacific. The species is not common and does not seem to be a nuisance to coral reefs like its larger relative. It is probably mildly venomous. Photo by Alex Kerstitch.

Detail of the spines of *Acanthaster ellisii.* Photo by Alex Kerstitch.

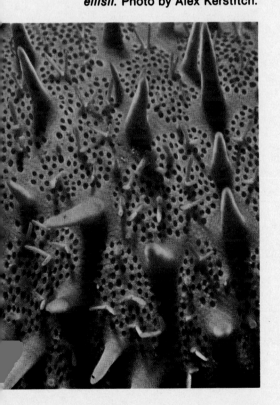

STARFISHES

In many starfishes there are heavy spines on the body that can cause local irritation and swelling if they are allowed to puncture the skin. Probably the worst culprit in this regard is the crown of thorns starfish, *Acanthaster plancii,* of the tropical Indo-Pacific. This large starfish is covered over the entire body and arms with large and small spines just under the skin. The skin is mildly venomous, so careless handling results in not only punctures from the spines but also mild envenomation. The pain is sharp and numbing, but it soon subsides, usually without aftereffects. The crown of thorns is more renowned as the starfish that is gradually destroying several reefs around Australia and Hawaii than as a venomous animal. Because of some poorly understood changes in the ecology of the reef, in certain areas the shrimp and other predators that normally control the size of *Acanthaster* populations are ineffective and literal "swarms" of hungry *Acanthaster* have moved over the reefs, feeding on the living coral polyps and leaving only dead reefs behind them. The best control so far seems to be using divers to locate the starfish (they are large and easily found where the ecological imbalance is present) and inject them with formalin, killing them on the spot. Fortunately, the stories of destroyed reefs are greatly exaggerated and what danger the starfish poses to the reef is usually considered minimal in most areas.

3: Molluscs

Snails, clams, squids, and their allies comprise the abundant and familiar phylum Mollusca. Many of these animals are instantly recognizable by their calcareous shell, though others lack a shell entirely or have it located internally. Clams lack a head and jaws (in the usual sense) but have become famous as accumulators of water-borne bacteria and toxins passed on to eaters of raw or poorly cooked seafood; as they are not venomous, they will not be discussed further. Two groups of molluscs, however, are definitely venomous in the worst sense of the word. These are certain species of cone shells and one or more species of small octopus.

CONES

Cones are marine snails (Gastropoda) belonging to the very large family Conidae. Of the over 300 species of cones, most are found in the tropical and subtropical waters of the Indo-Pacific, with lesser numbers of species in the tropical Atlantic. A few species extend into cooler waters such as the Mediterranean, southern California, and New Zealand. The shell is usually 1 to 4 inches long, rather simply conical in shape, and has a relatively large aperture; many species are grayish to brownish in color, but others are very brightly colored with interesting patterns. Because of the diversity of the shells, this family is very popular with shell collectors, some species fetching very high prices (as much as $2,000 to $4,000).

All cones feed in much the same way. The scraping radular teeth of most other snails have become modified into slender, barbed, hollow teeth that are stored in a sac behind the proboscis in the anterior part of the body. The teeth are indirectly connected by ducts to a large, usually kidney-shaped "venom bulb" that probably does not really secrete venom. The granular venom is apparently produced by the

This small pelagic nudibranch, *Glaucilla marginata,* has the ability to reuse nematocysts taken from its coelenterate food, especially *Velella.* Fortunately the small size of the nudibranch prevents it from being an unexpected danger. Photo by Dr. T. E. Thompson.

Facing page: Octopuses figure in many myths about the sea, with numerous stories emphasizing how giant octopuses (actually nonexistent) can kill swimmers and fishermen. Actually, all octopuses and squids are theoretically dangerous because of venomous saliva that can be introduced with a vicious bite. The bite of even a small squid or octopus can be very painful and cause local swelling.

Three dangerously venomous cones: **Above:** *Conus textile*, detail of the anterior end with the proboscis and one eye visible; there is still some doubt whether this species is truly venomous to man. **Below left:** Living animal of *Conus geographus*, one of the most deadly species. **Below right:** Living animal of *Conus striatus*, the other species most often involved in serious envenomations. Photo above by Keith Gillett from *The Australian Great Barrier Reef;* photos below by Alex Kerstitch.

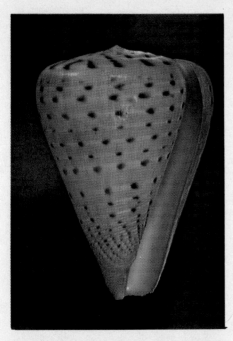

Ventral view of the shell of *Conus betulinus*, a large vermivorous species that seems to be harmless. Photo by Jerry G. Walls.

Ventral view of *Conus aulicus*, a large molluscivore that is reputed to cause serious human envenomations although laboratory tests seem to discount this. Photo by Jerry G. Walls.

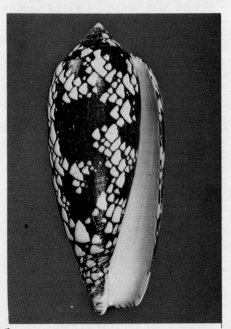

lining of the venom duct and fills the teeth as they mature. When a cone sights its prey, the long and flexible proboscis strikes out and literally spears the prey with one or more of the hollow teeth. The venom is potent and fast-acting, and the prey is dead or inactivated within seconds or minutes.

The venom of cones is apparently well adapted to certain types of prey. As a rule, cones fall into one of three categories of prey choice: some feed almost exclusively on worms, others on molluscs, and still others on fishes. The venom of worm-eaters (vermivores) is deadly to worms but has virtually no effect on vertebrates. Mollusc-eaters (molluscivores) have a more potent venom that at least in some species may be mildly toxic to vertebrates, including man. Piscivores (fish-eaters) have a venom that of course affects vertebrates most strongly, and it is among the piscivores that species deadly to man are found.

If a person is stung by a molluscivore such as *Conus textile* or its relatives, there may or may not be serious aftereffects. The sting of a piscivore such as *Conus geographus* or *Conus striatus* is serious and occasionally deadly. The stings of most other cones are usually no worse than a beesting. Because so many people are interested in collecting cones for shell collections, we will discuss them in some detail.

Some of the larger vermivores, such as *Conus betulinus* and *C. leopardus*, could theoretically be dangerous to man because of their large adult size and assumedly large amount of venom produced. Fortunately, their venom is almost ineffective against anything but worms, so the worst symptoms of a sting are just localized swelling and short-lived pain. Even smaller vermivores such as *C. ebraeus* pack a powerful sting, however.

There is still quite a bit of controversy as to whether or not the molluscivores are dangerous to man. There seems to be little doubt that stings by *Conus textile* and *C. aulicus* have caused human fatalities (although there is always a possibility of misidentification of any cone), but in laboratory experiments the venom of these species is virtually ineffective even against mice. It is hard to believe that a venom that causes a mouse minor discomfort could kill an adult human. Regardless, the molluscivores are relatively unaggressive species that are usually recognized by the brownish tented pattern easily seen through the thin periostracum (the layer covering the shell and often opaque in other cones). Because of the literature reports of human fatalities, it is best to be careful when handling tented cones, especially individuals over about 2 inches in length. You would not want to be the textbook example that proves their venomous status.

There is no doubt that the larger piscivores such as *Conus geographus* and *C. striatus* not only are capable of killing humans but also have done so in the past. These two species (and their close relatives) have shells with large apertures and are large, active animals with long, very flexible proboscises easily capable of reaching around to sting any human hand picking them up. The teeth can be almost half an inch long and strong enough to pierce thin cloth, so they are nothing to be triffled with. Bites from these two species cause at least temporary paralysis of the limbs and prolonged difficulty in breathing. Smaller piscivores such as *C. tulipa*, *C. magus*, and *C. catus* are also dangerous, but their smaller size probably eliminates them as possible sources of human fatalities (although their effects on a child might be more serious).

Conus spurius, a common Caribbean species, is rather unusual compared to most cones in that it seems to be an omnivore feeding on both molluscs and worms and possibly on fishes as well. It can inflict a very uncomfortable sting in humans.

The venom of the larger piscivores acts in much the same way as the venom of cobras and other elapid snakes, with only a very short time being available for treatment before the effects of the sting become virtually irreversible. There is immediate and localized pain followed by numbness. The venom stops nerve action in the extremities, producing in addition to numbness or a tingling sensation such symptoms as swelling, redness, dizziness, vomiting, and difficulty in speaking. The diaphragm is soon affected, causing difficulty in breathing. If total paralysis of the diaphragm occurs, death due to respiratory failure results in a few hours.

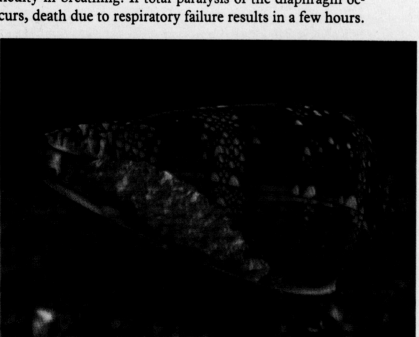

Ventral view of the shell of *Conus spurius,* an omnivorous Caribbean species that has caused uncomfortable stings when handled. Photo by Alex Kerstitch.

The East Pacific version of *Conus textile* is *Conus dalli,* shown here in ventral view with the animal retracted. It should be treated with caution. Photo by Alex Kerstitch.

There is no effective general treatment for cone stings. If the sting is from a truly deadly species, immediate hospitalization is essential. Otherwise the effects of the sting will disappear in a few hours or a few days, depending on the species and severity of the sting plus such factors as the health and size of the victim. The best first aid is simply to see a doctor if you have any doubts about just which species stung you. In fact, it is best never to handle a cone with a bare hand as it is sometimes almost impossible to identify a cone to species in its natural environment because of the heavy periostracum that obscures the color pattern in some species. Cones should be considered at least theoretically as dangerous as poisonous snakes and should be treated with the same respect.

OCTOPUSES

The cephalopod molluscs include the nautiluses, cuttlefishes, squids, and octopuses. All cephalopods have the head area (the foot in other molluscs) divided into a variable

The Australian blue-ringed octopus, *Hapalochlaena maculosa,* from southeastern Australian. The pattern of the rings varies considerably. Photo by Walter Deas.

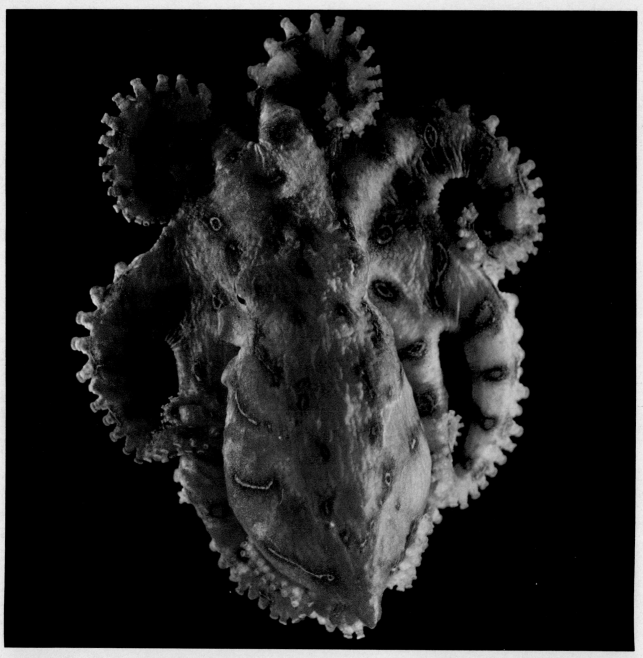

number of arms, usually eight in octopuses and ten in cuttlefishes and squids. The shell is coiled and external in the nautiluses, reduced to a horny pen or spongy internal shell in the squids and cuttlefishes, or absent in the octopuses. The cephalopods are carnivores feeding on fishes, crabs, and other marine animals, usually catching them with the sucker-bearing arms and then biting them with the parrot-like beak and injecting a potent saliva from large salivary glands. In theory, all cephalopods should be considered venomous as the saliva may cause at least local reactions at the site of the bite. Certainly the bite of the sharp beak is painful and not worth risking whether a venom is present or not.

Although it is less than 8 inches or so in maximum spread, the blue-ringed octopus is one of the most venomous marine animals. Photo by Keith Gillett from *Australian Seashores in Colour.*

All squids and octopuses have a mouth with beak-like jaws. The bite of the cuttlefish, *Sepia,* has caused local reactions among fishermen.

Common octopuses are unaggressive animals and are not likely to bite unless severely abused. The species commonly kept in aquaria (except the blue-ringed) are probably harmless. Photo by Dr. T. E. Thompson.

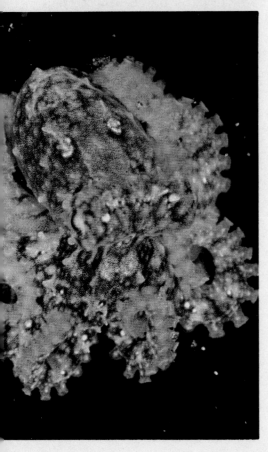

The only known dangerously venomous cephalopods are the small blue-ringed octopuses, *Hapalochlaena.* Two species are recognized: *H. maculosa,* from the subtropical and temperate coastal waters of eastern Australia, and *H. lunulata,* from the tropical waters of Australia north through the Indonesian and Philippine area. There is some question as to whether or not the northern *H. lunulata* is venomous as there seem to be no records of human biting incidents and local fishermen regard it as harmless. The two species are very similar, however, and both are best treated as being dangerous.

Both species are small (about 8 inches maximum spread) inshore octopuses that hide in crevices and debris, especially in clumps of sea squirts. Both are brownish to purplish with bright blue stripes that tend to form complete or nearly complete circles on the tentacles and more open crescents on the body. They can be very common in an area and often draw the attention of waders and divers because of their brilliant color pattern. Females carry the eggs in a cluster under the bases of the tentacles rather than attaching them to the substrate as in other octopuses. The animals become sexually mature very rapidly (within four months of hatching) and possibly die after spawning.

There are several records of human fatalities due to bites of *Hapalochlaena maculosa* on the southeastern Australian coast. In all cases the bites occurred after a wader or diver picked up a specimen of the octopus and placed it on the back of the hand to show it off to companions. The bite itself is painless and not detected until symptoms begin to occur. In five minutes or so after the bite the victim becomes dizzy and complains of difficulty in breathing. Respiratory paralysis rapidly sets in, and artificial respiration becomes necessary to sustain life. If prompt hospitalization is not available, the victim dies from complete respiratory failure within two or three hours of the bite.

Laboratory tests have shown that the venom of the blue-ringed octopus is actually a very potent saliva from the large posterior salivary glands. In some ways the saliva is as potent as the venom of the cobras, acting in the same neurotoxic fashion. As it apparently forms no antigens when injected into laboratory animals, antivenins are difficult or impossible to produce. The only treatment is immediate artificial respiration continued until the action of the poison ends, which may be several hours or even days. As the octopus is unaggressive, bites will occur only if the species is actually handled, so the safest procedure is to not handle blue-ringed octopuses and avoid trouble.

31

4: Worms

The group of animals commonly called worms is a very artificial one containing members of many different and unrelated phyla. Few worms are venomous, and the only ones likely to be found are a few polychaetes belonging to the phylum of segmented worms, Annelida. These are typically "worm-shaped" marine animals with many body segments, most of the segments having lateral extensions known as parapodia. The parapodia have muscles that are used to move variously shaped setae or bristles that are used in locomotion. The jaws are often strong and may even be able to draw blood. None of the species are dangerously venomous, but several types have irritating bristles or painful bites.

BRISTLEWORMS

Several groups of marine annelids have dense mats of fine bristles either covering the entire dorsal surface of the body or concentrated in tufts by the parapodia. The group as a whole is called the bristleworms; if the body is wider, almost oval, with longer and heavier spines among the bristles, the worms are sometimes called sea mice.

In the venomous members of this group the bristles are very fine, glass-like, and hollow. A fluid suspected of being the venom has been seen filling the centers of the bristles. The bristles readily penetrate the skin on contact, even at a very casual touch, and will go through gloves in some cases. They cause intense local pain, itching, and swelling that may last for several hours.

Although many polychaetes have stinging bristles, only four genera are commonly indicted as being especially dangerous. The large sea mice of the genus *Aphrodita* have not only the carpet of fine bristles covering the entire top of the body, but they also have heavy spines capable of causing considerable pain. Species of *Chloeia* and allied genera

One of the few types of worms that might be potentially dangerous to divers and aquarists is *Hermodice carunculata*, which is found on reefs in all tropical seas. It also sometimes appears in the aquarium trade. Photo by Dr. Dwayne Reed.

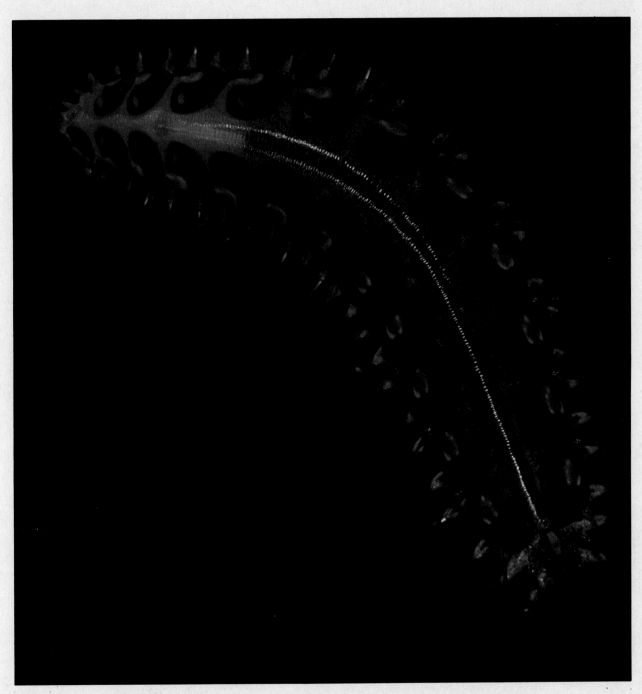

It is probably best to avoid handling any relatively large worm that has obvious spines or plates on the back that might have mats of fine bristles. Although few of the scaleworms are actually harmful to man, there are a great many species in the tropical waters of the world that could be potentially harmful. The worm featured here, *Lepidonotus melanogrammus,* is from Australian coastal waters and does not have a reputation for causing rashes when handled. Photo by Keith Gillett from *Australian Seashores in Colour.*

Above: Two views (ventral at left, dorsal at right) of an 8½-inch *Aphrodita* species from the East Pacific. A sea mouse this size can cause great discomfort if incorrectly handled. Photos by Alex Kerstitch. **Below:** The two most familiar bristleworms that are dangerous when handled are (left) *Eurythoe* sp., here represented by a 4-inch specimen from the East Pacific, and (right) *Hermodice carunculata,* a 10-inch worm from Caribbean reefs. Photo at the left by Alex Kerstitch; that at right by Dr. Patrick L. Colin.

This iridescent scaleworm of the family Polynoidae is probably harmless, but it still should be handled carefully. Photo by Scott Johnson.

are relatively stout and have clusters of bristles only on the sides of the body, with spots or stripes dorsally.

The really dangerous bristleworms are species of the genera *Eurythoe* and *Hermodice*. Both are long, slender worms with bright red gills above clusters of white spines. *Eurythoe* is up to about 6 inches long and is orange yellow in color; *Hermodice* may reach 10 inches in length and is greenish in color. Both genera are typical of reefs, with *Hermodice* found in the Caribbean and *Eurythoe* found in all tropical seas.

BITING WORMS

Many polychaete worms have strong jaws capable of inflicting painful bites. The common clamworms used for bait in many areas, *Nereis* and allies, have heavy retractable jaws that can suddenly pop out of the head and grab a probing finger in their sharp points, drawing blood. The only result of such a bite is usually just painful local swelling, and it would appear that venom is not associated with the bite. In the bloodworm, *Glycera*, the retractable proboscis ends in four sharp teeth that are equally spaced around the open edge of the proboscis. Each tooth is connected to a venom gland and is capable of producing toxic effects such as minor swelling and itching.

It is doubtful if medical attention will be needed for the stings or bites of any polychaete worms. Alcohol or a mixture of ammonia and alcohol can be applied to ease the burning sensation. Bristles should be removed as soon as possible; masking tape pulled over the point of contact usually works.

The species of *Nereis* and similar worms used for fish bait pose a minor threat. Photo by Ken Lucas, Steinhart Aquarium.

35

5: Fishes

Of the 20,000-30,000 species of fishes known so far, a great many have developed excellent defensive mechanisms such as heavy teeth, electric organs, strong and pointed fin spines, caudal fins capable of smartly slapping a predator, and various types of venomous stings. In many cases the stings are very well developed, associated with strong dorsal spines or pectoral spines, and capable of causing severe pain or even death in humans. Because of the large number of species that possess venomous stings, we will only be able to discuss the most important ones here. However, as a general rule (with many exceptions), it is best to assume that any fish—freshwater or marine—with heavy, pointed dorsal and pectoral fin spines or heavy spines alongside the head is dangerous. Certainly many fishes, even though nonvenomous, can inflict painful wounds with their spines and are best left alone or handled with extreme care.

Venomous fishes can be divided into two major categories: fishes that sting with a barbed caudal spine and fishes that sting with dorsal, pectoral, opercular, or other body spines.

STINGRAYS

Many species of sharks, rays, and chimeras have spines in the fins, scattered over the body, or on the tail. In several cases there are heavy spines at the front of the dorsal fin or fins, as in chimeras and in several shark genera including the common *Squalus*, and these have been indicted in severe human stings on occasion. However, these animals are likely to be handled only by professional fishermen working in rather deep water or certain coldwater ports and are not often seen by other people, so we will suggest only that any shark with dorsal spines be handled carefully or not at all. The elasmobranch fishes that concern us are the many species of rays capable of inflicting painful wounds with a barbed spine on the tail—the stingrays.

An unusually attractive pattern of the scorpionfish *Scorpaena neglecta*. The many species of this genus are usually cryptically colored. Photo by M. Goto, *Marine Life Documents*.

Facing page: *Taeniura lymma*, the blue-spotted stingray, is a common Indo-Pacific species that is also imported for the aquarium hobby. The long, muscular tail is quite effective in propelling the short, thick tail spine. Photo by Allan Power.

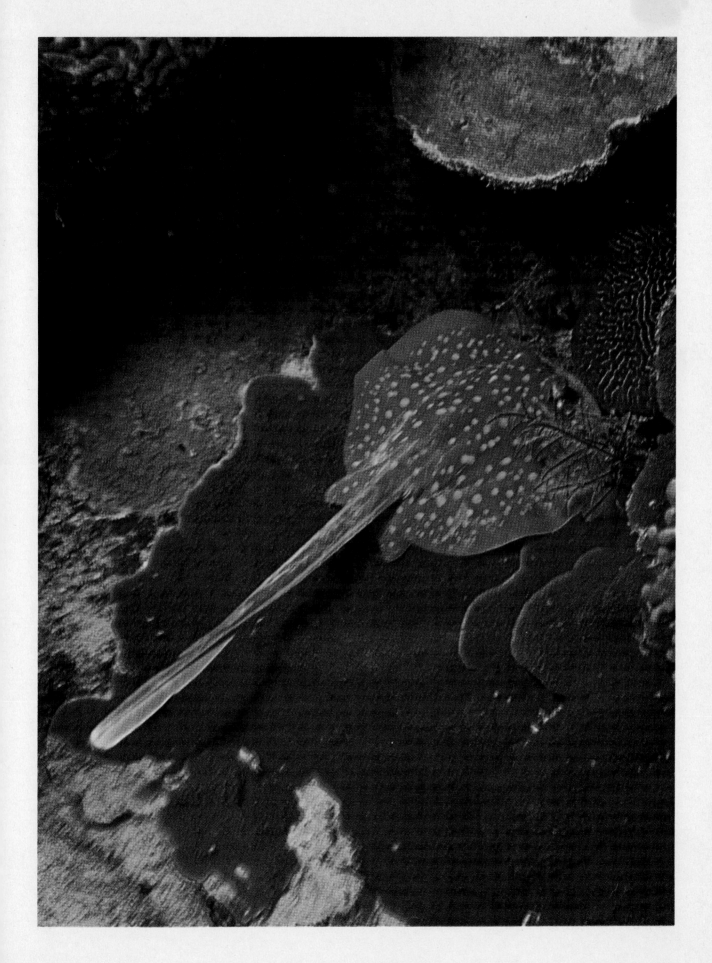

Skates and rays are basically flattened, disc- or diamond-shaped shark-like fishes with the gill slits on the ventral surface of the body and with the pectoral fins greatly expanded and usually attached to the head. The tail is usually long and narrow, with or without a small caudal fin. Most rays and skates are grubbers in the bottom mud, feeding on clams and invertebrates, but a few, such as the mantas, are pelagic and strain planktonic animals from the water.

Several families of rays have one or more spines attached somewhere along the length of the tail, but only four families are large enough and active enough to be considered dangerous. These are the Dasyatidae, the true (usually marine) stingrays; the Potamotrygonidae, the freshwater stingrays of tropical America (very closely related to the Dasyatidae and differing mainly in the structure of the pelvic girdles); the Urolophidae or round stingrays, small marine species with strongly muscled tails and heavy spines capable of causing severe wounds; and the Myliobatidae or eagle rays, large species with tail spines that may exceed 4 inches in length.

Amphotistius kuhli, an Indo-Pacific stingray. Photo by Heiko Bleher.

A typical *Dasyatis* species, one of the more dangerous types of stingrays and much like the species found on the coasts of the United States.

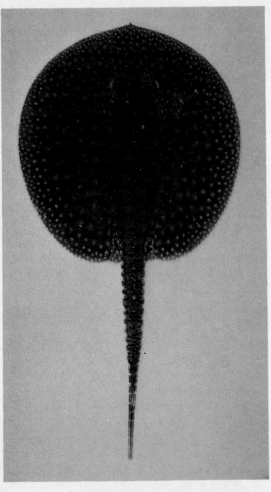

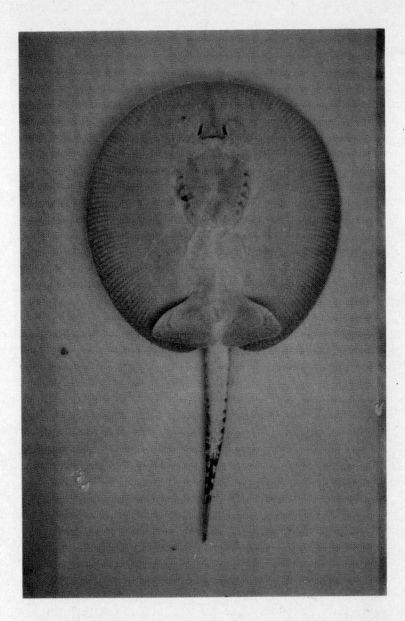

Dorsal (above) and ventral (right) views of *Potamotrygon reticulatus,* a typical South American freshwater stingray. The attractive color pattern and oddity of these fishes make them of great interest to aquarists, but they are definitely dangerous if improperly handled. Photos by Dr. Herbert R. Axelrod.

Urolophus aurantiacus, a small round stingray from Japan and the Indo-Pacific. Photo by Dr. Fujio Yasuda.

As mentioned earlier, rays are usually grubbers in bottom mud, where they pick up clams and other invertebrates and crush them on several series of flat teeth. They are commonly either completely buried in the mud or partially buried with just the top of the head and the tips of the pectoral "wings" showing. Stingrays are not aggressive animals, but when stepped on by the unwary wader or diver they react quickly, swinging the tail and its spine upward with sufficient force that the spine can actually pass through a man's leg or enter the bone. Fortunately, a stingray cannot effectively swing its tail unless there is a point of leverage—in this case the downward force of the wader's foot at the back of the ray's disc. Thus, in waters inhabited by stingrays waders are careful to "shuffle" their feet when in the water, never lifting them high enough to actually step on the ray and give it the required leverage.

Regardless of the family to which they belong, stingrays are characterized by the presence of a long, flat, barbed spine on the tail. The position and actual size of the spine vary considerably with family and species, and some species are more dangerous than others. In species with heavily muscled tails with the spine toward the end of the tail (as in some potamotrygonids and the urolophids) the spine can inflict extremely severe wounds, while in species with more slender tails and the spine more basally located the leverage is not as effective and the sting is thus not as dangerous. On the underside of the stinging spine are two or more shallow grooves containing strips of grayish tissue. It is this tissue that actually produces the venom of the stingray. The entire spine is usually covered with a thin layer of skin that is rubbed off the barbs when the spine enters the foot or leg of the victim, but even a spine without a skin covering still has the venom-producing tissue in the ventral grooves and thus remains dangerous.

The wound produced by a stingray can be either a puncture or a cut, but either way the results are similar. The wound is extremely painful, often with severe bleeding from the effect of the barbed spine being twisted by the ray as it is pulled out (a stingray may leave part of the spine in the wound, but this is exceptional). Rapid swelling follows. The pain may continue for several days, with cramps and nausea as well. Stingray wounds are notoriously slow to heal, and secondary infections or even gangrene may rapidly develop. If not given prompt and continuing medical attention, it is not unusual for the wound to result in loss of the foot or the entire limb. Deaths are not rare, though they are most often the result of the secondary infections. In several instances divers have landed on the back of large stingrays and received the spine in the chest cavity, often with fatal results.

Serious stingray wounds require immediate medical attention, as stitches and antibiotics are often necessary. First, however, clean the wound as thoroughly as possible in the field, making sure that any broken pieces of spine are removed and that no tissue from the surface of the spine (including venom-producing tissue) is embedded in the wound. Many authorities recommend that the wounded limb next be placed in very hot water—as hot as the victim can stand without blistering—for 30 to 90 minutes to help deactivate the venom. At the very least, however, it is necessary to clean out the wound, get medical attention, and be prepared to deal with secondary infections. These same steps can be said to apply to the sting of almost any other venomous fish, from a shark to a lionfish.

Detail of the tip of the tail of a *Potamotrygon* species. Notice that the spine is virtually at the end of the tail, increasing the leverage that can be applied and making the sting more dangerous. Photo by Andre Roth.

Urolophus jamaicensis, the common yellow stingray of the Caribbean. This relatively small species is a dangerous stinger. Photo by U. E. Friese.

Even in an aquarium this *Potamotrygon* is able to camouflage itself among the gravel, so you can imagine how effectively it can disappear into the substrate of its native rivers and streams. The risk of stepping on a hidden stingray is great wherever they occur. Photo by Andre Roth.

41

Noturus insignis, the margined madtom, is a common species from the eastern United States. These largely nocturnal fish are seldom noticed by fishermen or others and are unlikely to cause any harm unless handled. Photo by Dr. Lawrence M. Page.

Much more complexly marked is *Noturus flavater,* a species from the central United States. Some species of madtoms have very heavily barbed pectoral spines and could theoretically cause considerable damage, but their small size keeps them from being considered really dangerous. Photo by Dr. Brooks M. Burr.

One of the catfishes seen most commonly in aquaria is *Pimelodus pictus,* a South American species called the angelicus pimelodus. Although this species (which seldom grows to over 4 inches in length) is not known to have caused noticeable human stings, it should be considered potentially dangerous and treated with respect. Photo by Dr. Herbert R. Axelrod.

STINGING CATFISHES

Everyone recognizes a catfish on sight—they are small to very large fishes with naked skin (plates maybe present but not true scales), barbels at the front of the head, and a single dorsal fin often followed by an adipose fin (a non-rayed fin-like structure filled with fat). There are many families of catfishes found all over the world, with several of the numerous species important commercially as foodfishes. In many of the species the spine at the front of the dorsal fin and the spines in the pectoral fins are very heavy, barbed, and covered with venom-producing tissue. These spines are often hollow and have special processes at their bases that allow them to be "locked" into an erect position. When handled, as when they are removed from a net or from a hook, the spines go up, lock in position, and become formidable weapons.

Catfishes are probably the most common and widely distributed of the at least theoretically dangerous fishes. Several species of North American ictalurid catfishes, especially the small *Noturus* species known as madtoms, are capable of inflicting painful wounds about equal in effect to a bad beesting. In tropical American rivers and streams there are numerous species of the genera *Pimelodus* and *Rhamdia* with equally or more potent stings, and many other South American catfishes are capable of stinging also;

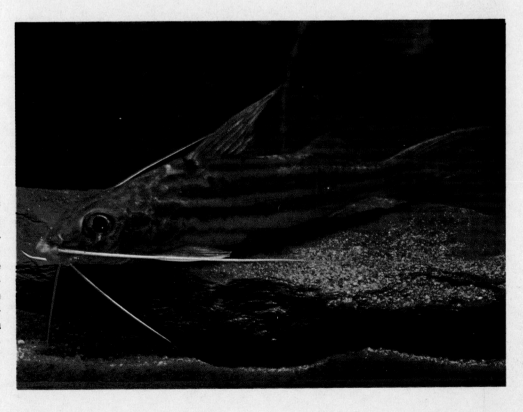

Large species of *Pimelodus*, such as this *P. clarias*, are of commercial importance in South America and are often involved in accidents. On occasion human deaths have occurred. Photo by Ruda Zukal.

When kept in aquaria and carefully handled, madtoms such as this *Noturus miurus* are harmless. Photo by R. L. Mayden.

North American food catfishes, such as the channel catfish, *Ictalurus punctatus,* are often roughly treated by fishermen yet seldom cause any serious stings from jabs of the fin spines. Photo by G. Purvis.

The white catfish, *Ictalurus catus,* is another large North American catfish that has the physical equipment to cause damage but is seldom involved in accidental envenomations. Photo by Ken Lucas at Steinhart Aquarium.

Close-up of the front part of the body of a *Clarias* species showing the large pectoral spine that can cause severe envenomations. Photo by Ruda Zukal.

Pimelodus albofasciatus is one of the larger commercial catfishes in northern South America. When large numbers of these fish become entangled in nets, it may take hours to remove them, and envenomations become likely or even unavoidable. Photo by Dr. J. J. Hoedemann.

The attractive pattern of *Pimelodus ornatus* makes it a very desirable aquarium fish. As is the case in other aquarium catfishes, there are virtually no records of serious envenomations caused by captive animals, regardless of how potent the fish may be in nature. Photo by Dr. Warren E. Burgess.

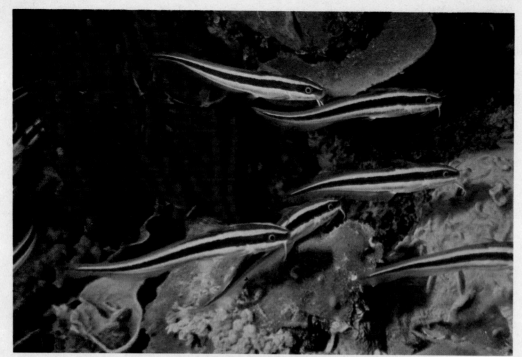

Brightly striped specimens of a few *Plotosus* species are common and desirable marine aquarium fishes, but the adults have the reputation of being very serious stingers. Even the young can sometimes inflict a painful sting. Photo by Roger Steene.

stings from *Pimelodus* have resulted in human deaths. The marine catfishes belonging to the family Ariidae, including the common sea catfish and gaff-topsail catfish of the Atlantic Seaboard and Gulf of Mexico, have often been indicted in human stings that are sometimes serious. In this family the spines are often hollow and covered with large masses of venom-producing tissue. Ariid catfishes are found in coastal and estuarine waters of the Americas, Africa, Australia, and Asia and are often extremely abundant both in number of species and number of individuals. The Asian and African Clariidae contains elongated catfishes with a long dorsal fin often connected to the caudal fin. There are many species, of which the common *Clarias batrachus* or walking catfish is a familiar foodfish in Asia and has been introduced and escaped in southern Florida. Clariids have well-developed pectoral fin spines capable of producing a dangerous sting.

Probably the most dangerous catfishes belong to two Asian families. The freshwater *Heteropneustes fossilis* (Heteropneustidae) is found from India to Vietnam. Although small in size, its spines are capable of inflicting painful stings that have taken human lives on several occasions. Unlike most catfishes that sting only when mishandled, *Heteropneustes* has the reputation of actually attacking intruders into its waters and voluntarily stinging. It is an elongated catfish with a small dorsal fin and long anal fin.

The marine and estuarine Plotsidae are common in southern Asia and from Australia to eastern Africa. Although one or two species of *Plotosus* are common aquar-

Because of the reputation *Clarias batrachus* has gained in Florida, importations of all *Clarias* species are at least closely scrutinized when they enter the United States and some other countries. Photo by Dr. Herbert R. Axelrod.

It's hard to believe that this foot-long catfish, *Heteropneustes fossilis,* could be one of the most deadly of fishes, but it seems to be true. Although abundant and edible, it is avoided by fishermen throughout its Indian and Southeast Asian range because of its poisonous spines. Photo by Edward C. Taylor.

The heavily armored doradid catfishes are commercially important both as aquarium fishes when small and as food fishes when large. The odds are that small aquarium specimens such as this *Amblydoras hancocki* are harmless to humans. Photo by Ruda Zukal.

ium fishes, both adults and juveniles have potent venom and sting freely. Adults often cause human fatalities, especially among fishermen clearing their nets. The adults of aquarium *Plotosus* species are probably just as dangerous as lionfishes and should be handled with great care.

In all the catfishes the stinging apparatus is similar. It consists of the strong and often barbed and hollow fin spines and large or small masses of venom-producing tissue on the surface of the spine. When the spine penetrates the victim's hand or foot, the thin layer of skin over the spine is abraded and torn, exposing the venom tissue. The barbs produce a relatively deep wound, allowing the venom to rapidly enter the system. Although the spines of many catfishes are hollow, there is no venom gland at their base and the venom is not actually injected, as one might first think. Often the sting produces only local swelling and a stabbing pain that may last just a few minutes, but in more severe cases there may be muscle spasms, extensive swelling, and difficulty in breathing. As mentioned, there have been human fatalities from the stings of several different catfishes, so any catfish encounter should be watched carefully and brought to medical attention if unusual symptoms develop.

LIONFISHES AND OTHER SCORPIONFISHES

The large family Scorpaenidae consists of several types of marine (and occasionally estuarine) fishes that are especially prominent on tropical reefs. Scorpionfishes derive both their common and scientific names from the resemblance of the effects of their stings to the well-known stings of scorpions. Probably every scorpionfish is venomous to some degree or another, and unfortunately the most familiar aquarium species are among the most venomous.

It must be stated in passing that scorpionfishes are often broken into several families on morphological grounds, with *Scorpaena* and allies placed in the Scorpaenidae proper, *Synanceia* and relatives put in the Synanceiidae (stonefishes), and *Pterois*, *Brachirus*, and allies put in the Pteroiidae (lionfishes). Among American taxonomists the tendency is to recognize only two of these families, Scorpaenidae (including Pteroiidae) and Synanceiidae, but some taxonomists recognize as many as four or five families, plus two or three more for some unusual scorpionfish allies such as the velvetfishes. At least in theory, all these fishes could be venomous regardless of their systematic placement.

It is unfortunate that one group of these fishes, the lionfishes, includes several species that are extremely popular and attractive aquarium fishes. These are members of the genera *Pterois* (*P. antennata*, *P. lunulata*, *P. volitans*, etc.)

One look at this spot-finned lionfish, *Pterois antennata,* is enough to show why they are so popular in the aquarium hobby. Photo by H. Hansen, Aquarium Berlin.

Even juveniles of the less spectacular lionfishes, such as this *Brachirus barberi,* are commonly imported for the pet trade. Photo by Scott Johnson.

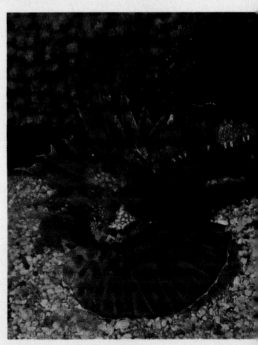

In the lionfish genus *Brachirus* (here represented by *B. biocellatus*) the rays of the pectoral fins extend little or not at all beyond the membrane. Photo by K. H. Choo.

and *Brachirus* (*B. biocellatus, B. brachypterus, B. zebra,* etc.), commonly known by such names as turkeyfishes and zebrafishes as well as lionfishes. The species are usually 4 to 8 inches long, have large heads with big mouths, are rather brightly banded in reddish brown and white, and have large, filamentous dorsal and pectoral fins. They are reef-dwellers that are often found upside-down in coral caves or other shelters. When annoyed they tend to stand their ground and may actually approach the intruder with the head down and the long dorsal spines erected. This display of confidence by the fish is not without cause, as their dorsal spines, anal spines, and pelvic spines all bear venom glands.

If someone is so unlucky as actually to grasp a lionfish or accidentally brush his hand against a cornered specimen, the reaction is instantaneous. First there is a burning, searing pain of high intensity. This is followed by swelling, often to a marked degree, and occasionally by such alarming symptoms as convulsions and cardiac arrest. The intense pain may continue for hours, causing disabling of the victim and even unconsciousness. Secondary infection is common, and gangrene is not unheard of. Recovery is often

As can be seen in the photo above of *Pterois volitans,* the dorsal spines when erected slant alternately to the left and the right. Shown below is *Pterois lunulata,* one of the more venomous species of the family. Photo above by Walter Deas; that below by K. H. Choo.

a slow and painful process of weeks or months, and there may be a permanent scar. Certainly these attractive fishes are not to be taken lightly.

The actual envenomation process is very simple. Each spine in the fins has on its surface an elongated venom gland covered by skin. When the spine penetrates the victim's skin, the sheath is pushed back and the venom gland empties part of its contents into the wound. The severity of a sting is thus at least partially dependent on the number of spines that actually penetrate the skin deeply enough to cause the glands to function.

Any sting by a *Pterois* or *Brachirus* should be considered possibly dangerous and medical attention should be sought. The same is also true for any sting from the various other scorpionfishes (including the common *Scorpaena*) that causes more than just a local and temporary reaction. A hot water treatment like that for stingray wounds has been suggested by some authorities, but with stings from several lionfish spines this would at best be a very temporary measure until medical attention could be obtained.

A *Pterois* reacting to a challenge is quite an impressive sight. The large mouth is opened widely and all the fins are broadly flared. Photo by M. Goto, *Marine Life Documents*.

Camouflaged scorpionfishes are often impossible to see in their natural habitat. Here an almost perfectly hidden *Scorpaena verrucosa* shows bright red inner surfaces of the pectoral fins, flash colors used to warn possible intruders. Photo by Scott Johnson.

The shores of the North Pacific have their share of scorpionfishes, most of them belonging to the genus *Sebastes*. These often large fishes are sometimes of great commercial importance as sportsfishes and foodfishes. Fishermen report occasional serious effects from stingings by these fishes. Photo of *Sebastes nebulosus* by Al Engasser.

The plainly colored *Scorpaenodes scaber* from southeastern Australia is a typical scorpionfish much like the species common in all warm and temperate oceans. Photo by Rudie Kuiter.

The systematics of scorpionfishes is based largely on the number, position, and development of the spines on the head. As can be seen in this photo of an unidentified Pacific scorpionfish, the spines usually are numerous and can inflict serious cuts if the fish is carelessly handled. Photo by Allan Power.

STONEFISHES

Although the stonefishes of the family Synanceiidae are closely related to the Scorpaenidae (so closely that some taxonomists do not recognize them as a separate family), their form is quite different from that of the lionfishes. These are chunky, usually depressed fishes with large and usually upturned mouths. They are sedentary, usually lying partially buried in the debris of a coral reef or in mud flats, waiting for prey to come close enough to be gulped down. Their colors are usually at least subdued, more often ugly, matching to some extent their background of algae and debris. There often are filaments and skin processes that help further hide their form.

In most members of this group the venom glands are very similar to that of *Pterois* and other scorpionfishes, but in the fishes of the genus *Synanceia* the spines and glands are exceptionally well developed. Here the venom glands are rounded, are extremely large for the size of the spine, and are easily visible to the naked eye when the skin sheath is removed. The

Close-up of a stonefish, *Synanceia verrucosa*. This seemingly formless mass of flesh is indeed a large, exceedingly venomous fish and not just a random pile of debris on the ocean floor. Photo by Allan Power.

Seen from the side, *Synanceia* looks more like a chubby scorpionfish, which to some extent is just what it is. Photo by Dr. Herbert R. Axelrod.

Inimicus filamentosus is also considered to be a stonefish, although it admittedly bears little resemblance to *Synanceia.* The unusual shape of the head is found in many of its relatives. *Inimicus* is one of the more venomous scorpionfishes. Photo by Dr. D. Terver, Nancy Aquarium.

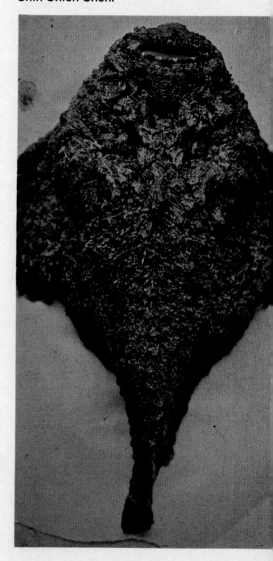

gland attenuates into a hollow duct that continues to near the end of the spine and has a distinct opening.

Synanceia lies partially buried in shallow water, its color and skin texture closely matching the background. The dorsal fin spines are erected at the least disturbance, so the unlucky wader or diver who steps on one of these fishes is immediately stung by the spines. If only one spine penetrates the foot, enough venom can be injected to cause convulsions and unconsciousness along with incredible pain and even paralysis of the limb. If the venom from several spines is injected, death can rapidly follow. At best, recovery from a less severe stonefish sting may take several weeks and have persistent aftereffects that last for months.

Stonefishes seem to be especially prevalent on the northeastern Australian coast, where there have been several deaths from unfortunate encounters. In coastal Australian hospitals a stonefish antivenin is available that is very effective in halting the effects of stonefish stings, but in more remote tropical localities antivenin is likely to be unobtainable. Stonefish spines can easily penetrate flippers and thin tennis shoes, as well as gloves, so it is best to wade cautiously in stonefish country and never handle these fishes under any circumstances.

OTHER VENOMOUS FISHES

Several other groups of fishes have one or more venomous genera, but usually they are of restricted distribution or are unlikely to be encountered by the average person. Among these are a few toadfishes, the

Most toadfishes, whether freshwater or marine, harmless or venomous, look very much alike. Shown is the Southeast Asian *Halophryne trispinosus*, a solid-spined (and thus harmless) species that enters fresh water and is sometimes kept in the aquarium. Photo by Aaron Norman.

weevers, some stargazers, rabbitfishes, scats, and perhaps tangs or surgeonfishes.

Tropical American toadfishes of the genera *Daector* and *Thalassophryne* (family Batrachoididae) are sometimes common in marine, estuarine, and freshwater habitats. These are bottom-dwelling fishes with two hollow dorsal fin spines and a hollow opercular spine on each side of the head. At the base of each of these spines is a large, rather conical venom gland. If the fish is stepped on or handled, venom is injected into the wound produced by the puncture. Usually toadfish stings are not very serious, although they are painful and there can be complications from secondary infections.

Weevers, genus *Trachinus*, family Trachinidae, comprise only a handful of species in cooler waters from Europe to western Africa. They are often common in shallow sandy bays and enter the foodfish trade in some localities. The solid dorsal fin spines and opercular spines have associated venom glands much as in the scorpionfishes and toadfishes. Stings from these spines occur during handling of specimens in nets or when preparing them for market; waders may also occasionally step on living specimens partially buried in the bottom. The sting of a weever is serious and may cause death. Even in less severe cases the pain may last for several hours, and secondary infections and gangrene

Above: All stargazers are similar in general appearance, with a chubby body, up-turned mouth, and dorsally directed eyes. This is the Chinese *Ichthyoscopus lebeck,* a 10-inch fish sometimes taken by commercial trawlers. All stargazers should be handled carefully as little is known of the potential toxicity of stings from their spines. Photo by Dr. Shih-Chieh Shen. **Below:** A large school of the rabbitfish *Siganus lineatus.* Photo by Allan Power.

Bites from moray eels have long been considered to be venomous by divers and fishermen, but there is little evidence to indicate that the effects of a bite are due to more than secondary infections. No one in their right mind would mishandle a moray, however, whether it is venomous or not. Shown is *Muraena griseus*. Photo by H. Hansen, Aquarium Berlin.

may develop to cause complications. In Europe, weevers are popular foodfishes in some areas, and accidents in handling dead market specimens have been reported.

Stargazers of the genus *Uranoscopus* (Uranoscopidae) are cool-water bottom-dwellers of northern seas. They lack dorsal fin spines or strong opercular spines but instead have a strong, projecting cleithral spine on each side with a conical venom gland around its base (the cleithrum is one of the bones that forms the skeleton of the pelvic fin girdle). If the fish is handled, as when it is taken from a net or off a hook, the cleithral spines penetrate the skin of the victim and introduce venom, much as in the other stinging fishes discussed. There are small fisheries for these stargazers in the Mediterranean and Japan. Fatalities have resulted from stargazer stings. Other genera of stargazers have short but strong dorsal spines (*Astroscopus*) or variously developed cleithral and opercular spines (*Gnathagnus, Kathetostoma*), but it is not known if they are truly venomous; several stargazers have electric organs in the head region.

Rabbitfishes (Siganidae) have numerous dorsal and anal spines that are sharp and strong, plus two spines in each pelvic fin. Each spine has associated with it a small venom gland somewhat like that of a scorpionfish and capable of delivering a painful sting. Usually the effects of the sting are short-lived and not much worse than a beesting. Species of *Siganus* are cultured for food in the tropical Indo-Pacific, and species of both *Siganus* and *Lo* are not uncommon in the aquarium trade.

Rabbitfishes are usually considered to be virtually harmless by aquarists, but their spines do have associated venom glands and could cause local swelling and pain. Photo of *Siganus punctatus* by Scott Johnson.

Striking color patterns and unusual shapes make the species of *Lo* very desirable aquarium fishes. Every aquarist realizes that the many strong spines could cause serious wounds if the fish is mishandled, even if there were not a weak venom associated with stings. Photos of *Lo vulpinus* (at left) and *Lo magnificus* (below) by Dr. D. Terver, Nancy Aquarium.

In most dangerous fishes the adults are significantly more venomous than the juveniles, but in the scats this tendency is reversed. Adult scats (right) are harmless (although the strong spines can cause considerable pain if the fish is treated roughly), while juveniles (below) have weak but definitely venomous stings. Few aquarists realize that juvenile scats (the stage usually kept in the aquarium) are potentially harmful. Photos of *Scatophagus argus* by Dr. Herbert R. Axelrod (right) and G. E. Schmida (below).

Although the dorsal spines of surgeonfishes are known to have small and probably ineffective venom glands, most people handling these fishes worry about the other end of the fish. The caudal peduncles of acanthurids have either a single movable spine (as in *Acanthurus leucosternon*, above) or two or more immovable spines (as in *Naso lituratus,* below) that are blade-like, very sharp, and can inflict deep cuts or slashes on an unprotected hand. Experts still disagree as to whether or not venom is associated with these blades, but the cuts are painful and easily infected. Photos by Dr. D. Terver, Nancy Aquarium.

Close-up of the pair of fixed blades on the caudal peduncle of *Naso lituratus.* Photo by Dr. Herbert R. Axelrod.

Juvenile scats (Scatophagidae) are common aquarium fishes (*Scatophagus argus* is the most common species) that are able to adapt to both fresh and salt water. Adults are marine to estuarine in habitat and are silvery fishes sometimes used locally for food in the tropical Indo-Pacific; they have a reputation as scavengers and are often found near sewage outlets. In juveniles there are small venom glands associated with the strong dorsal, anal, and pelvic spines. These spines can cause a mild sting with some pain and local swelling. Usually there are no bad aftereffects. As the fish grows, the glands become relatively smaller; in adults they are absent or ineffective.

Tangs or surgeonfishes (Acanthuridae) form a rather large family of mostly herbivorous fishes of tropical oceans. They tend to graze on algae on tropical reefs, often forming large schools. Many are very attractively patterned and colored, and they are popular aquarium animals. The strong dorsal spines have small venom glands capable of inflicting minor stings, but the major defensive weapon is a pair (occasionally several pairs) of blade-like bones, often movable, on the sides of the caudal peduncle in front of the tail fin. In some species these blades can be opened or closed much like a pocketknife. The blades are strong and sharp and can cause deep lacerations in anyone who tries to handle one of these fishes incorrectly. There is usually a great deal of pain and some swelling associated with cuts from the blades, but it is still uncertain whether there is an actual venom present or not.

6: Introduction to Arthropods

It is often estimated that there are over one million living species of animals. Of this million, probably fewer than 50,000 are vertebrates (fishes, snakes, birds, mammals, etc.), 200,000 or so are invertebrates that live in the sea (such as the molluscs, jellyfishes, crabs, and many others), and the majority of the remaining 750,000 or so species are insects, mites, and spiders. The large phylum Arthropoda, which contains the spiders, mites, and insects as well as the crustaceans, is characterized by having jointed, segmented legs. The Arachnida, which contains the spiders, scorpions, mites, and several other more obscure groups, have four pairs of legs, no antennae, and no true jaws. The Insecta, the insects, have three pairs of legs, one pair of antennae, and true jaws (that may be modified into sucking mouthparts); additionally, the majority of insects have wings in the adult stage. The insects and arachnids are almost all terrestrial or freshwater animals, with only a few species occurring in the seas. In contrast, the crustaceans (Crustacea), arthropods with usually five or more pairs of legs, two pairs of antennae, and gills, are almost all aquatic, with most crabs and shrimp living in salt water. Curiously, no crabs or shrimp are venomous, and even the parasitic copepods seldom cause problems. The next three sections will deal with the venomous terrestrial arthropods: the scorpions, the spiders, and lastly (but most importantly) the insects.

Although insects are the most abundant and dangerous of the terrestrial venomous animals (causing over half the deaths and a great majority of the serious cases of animal stings or bites in the U.S. each year) and spiders and scorpions are widely known and feared, two other minor groups of arthropods contain venomous species that are worth a casual mention.

Although almost no common species are dangerously venomous, spiders are still the most greatly feared arthropods. When they reach large sizes, such as this 2-inch silk spider, *Nephila clavipes*, almost any casual observer is sure to assume they are deadly venomous species, regardless of the facts. Photo by Dr. Sherman A. Minton.

Facing page: The biting flies and mosquitoes are just nuisances to most people, but to some they might actually be dangerously venomous. Arthropod salivas have the ability to cause different reactions in different people, and there are instances of human deaths from the bites of even the most innocuous insects.

Although not technically venomous (they do not inject their poisonous chemicals but can spray them), certain beetles are definitely dangerous to handle. The melodids such as the Spanish fly (below), *Lytta vesicatoria,* and even less famous beetles such as the bright red cardinal beetle (left), *Cissites,* possess a chemical called cantharidin that is highly irritating to human skin. If a beetle is thoughtlessly crushed or (in some species) the chemical is sprayed onto the skin, painful blisters result that may last for weeks and may become infected. Melodid beetles are found almost everywhere but are seldom common.

CENTIPEDES

The centipedes, class Chilopoda, are elongated, many-segmented arthropods that are commonly found under stones, in piles of wood, and in damp places; one genus is a common house pest even in the northern United States and is found over much of the world through introductions from its original Eurasian home. Most of the several hundred larger species are found in tropical areas of the world, with such species as *Scolopendra subspinipes* reaching over 10 inches in length on occasion and the southern U.S. *Scolopendra heros* often exceeding 6 inches. Most species are only 1 to 3 inches long, however.

Centipedes are instantly recognizable by their elongated, rather parallel-sided and flattened bodies with many similar body segments and many legs. In centipedes the segments usually number from 15 to as many as 170 (usually the most segments occur in small, thread-thin species), with a pair of long walking legs on almost every segment. Thus a centipede may have from 30 to over 300 legs depending on the species. Most of the larger and potentially dangerous tropical species have about 20 segments and 21 or 23 pairs of legs. The common house centipede, *Scutigera*, has 15 pairs. The head is covered with a solid one-piece shield, there may be small ocelli visible at the side or top of the head, and the antennae are long and many-segmented. Under the head are a pair of maxillipeds that are hook-like in shape and also in function, serving to grasp prey. In the base of each maxilliped is a venom gland connected by a duct to a small opening near the tip of the claw ("fang") of the maxilliped, so venom can be injected into the prey while it is being held. Centipedes usually feed on small living but slow-moving insects and occasionally on earthworms, thus the venom is adapted to work most effectively on cold-blooded prey.

When a human accidentally picks up a large centipede with a piece of wood or debris or if a centipede should happen to wander into a shoe or a piece of clothing, a bite may very easily occur as centipedes tend to be short-tempered, biting first and then trying to escape. If the centipede is large enough, the bite results in immediate severe pain that may last for several hours, plus local redness and swelling. The bites of large *Scolopendra* species are seldom serious, and deaths are almost unheard of.

A large house centipede has a very potent venom for its size (usually less than 3 inches body length), and the species may be abundant enough in human habitations to be a minor danger to pets and small children. *Scutigera* is readily recognized by the extremely elongate and thin legs that are

Detail of the anterior body of a large *Scolopendra* centipede. Notice the heavy maxillipeds ("fangs") under the head shield.

strongly "elbowed" when at rest. The animal moves very swiftly across walls and ceilings and is hard to hit. It should not be swatted with bare hands unless you are willing to take a chance on being bitten and suffering for a few hours. Since they eat flies and other household pests and are not aggressive, they are best avoided and tolerated.

TICKS

Ticks are familiar to any dog-owner and are abundant almost everywhere. They are actually very large mites that live by feeding on the blood of warm-blooded animals, including man. Although ticks are considered important carriers of several diseases, they are known to cause only one common type of toxic reaction, tick paralysis.

When a tick bites its prey, its specialized mouthparts are buried in the flesh and saliva containing an anticoagulant and other chemicals is injected. In most cases of tick paralysis a female tick is involved and the attachment site is on or near the head. The female tick must have a blood meal before laying eggs, and she commonly engorges with blood until several times her normal size before leaving the host. During this period small amounts of saliva are injected into the host to assist in feeding. If the saliva contains a chemical that blocks nerve messages to the limb muscles, a gradual state of paralysis results. The first symptoms include a general lethargy, followed by weakness, then by discoordination and eventual paralysis. The paralysis lasts only as long as the tick is on the host and feeding, so careful removal of the parasite results in rapid

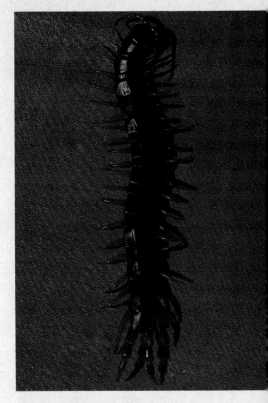

Centipedes, such as the large *Scolopendra viridicornis* shown above, are readily told from the harmless millipedes (such as the large African millipede shown below) by having only one pair of legs (at most) per segment, while millipedes have two, at least at the middle of the body. Photo above by Pedro Antonio Federsoni.

Species of *Amblyomma,* as well as other ticks, are intermediate hosts and vectors for many disease organisms and can occasionally cause tick paralysis directly. Photo by Dr. E. Elkan.

recovery with few side-effects. On rare occasions death has occurred, however.

Although many species of ticks have been associated with bites causing paralysis, the most common culprit in the United States is *Dermacentor andersoni,* a wood tick also known as the main carrier of Rocky Mountain spotted fever, a rickettsial disease of man. The common dog tick, *D. variabilis,* is also known to cause the paralysis, as are several species of *Amblyomma.* It is quite likely that most larger ticks that occasionally attack man have a saliva capable of producing tick paralysis. In Australia, for instance, *Ixodes holocyclus,* a species normally found on marsupials, may cause it.

Fortunately the condition is usually easily controlled by removing the tick. This must be done in such a way as to cause the tick to withdraw without leaving the mouthparts in the body of the host. The usual method is to apply petroleum jelly or fingernail polish to the entire body of the tick, causing it to suffocate because the spiracles are covered. Ammonia may also be applied, as it is enough of an irritant to sometimes cause the tick to free itself. The tick should be killed either by burning it with a match or by immersion in alcohol or ammonia. (Some cases of Rocky Mountain spotted fever are believed to have resulted from crushing an infected tick between the fingers.) The wound is then treated to prevent any secondary infection.

7: Scorpions

Next to spiders, scorpions may be the most vilified of the arthropods. Their distinctive appearance, wide distribution, and well-known toxicity have combined to make them both readily recognized and readily disliked. They were among the first, if not *the* first, arthropods to leave the water and adapt to terrestrial habitats, and they have no truly close relatives among the living arachnids.

Over 1500 species of scorpions have been described, most of them coming from the tropical and subtropical regions of the earth. Although popular conceptions of scorpions always present them in desert habitats, this is not true for the group as a whole. The common species of the western United States and Mexico, plus many African species, are indeed desert-dwellers, but a great many scorpions occur in savannah habitats or even in humid tropical woodlands; many scorpions are quite adept at climbing trees. Almost all species are active at night or at dusk, hiding by day in burrows or crevices. All species as far as known are carnivores, catching insects and even small vertebrates with the large, pincer-like pedipalps and stinging the prey with the aculeus or sting on the telson (the last segment of the abdomen). The prey is then transferred to the chelicerae, small grasping mouthparts under the front end of the carapace or head of the scorpion, where it is torn and covered with digestive fluids. It is further mangled by the bases of the front legs until it is flexible enough and liquid enough to be transferred to the mouth. This rather messy method of feeding has not helped endear scorpions to humans, who prefer to see jaws that move up and down, not semiliquid masses of prey held outside the body.

All scorpions are much alike in appearance, with no greatly divergent members of the order known. All have large pedipalps, relatively smaller chelicerae, a large flat carapace covering the head and thorax, four pairs of walking legs ending in claws, an abdomen of seven broad

Scorpions are among the most distinctive animals. All look very much alike in general form, and even throughout the fossil record scorpions have always been scorpions (although the early fossil scorpions were aquatic, believe it or not). These photos of a typical scorpion (above and facing page) show the characteristics of the group: the long posterior abdominal segments ending in the bulb and sting, the pincer-like pedipalps, the inconspicuous chelicerae under the front edge of the carapace, and the paired eyes plus ocelli on the carapace.

anterior segments and five narrow elongated posterior segments, and a rather bulbous telson at the end of the abdomen ending in a long, curved sting or aculeus. The telson contains two large venom glands surrounded by thick muscle and connected by fine ducts to two openings near the tip of the aculeus. Usually two large eyes are present near the anterior midline of the carapace, with a group of two to five small ocelli on each side at the lower anterior border of the carapace; the eyes are occasionally absent, especially in cave-dwelling scorpions, and the number of small ocelli varies considerably among genera. All the limbs and the pedipalps of scorpions have long sensory hairs called trichobothria that are important in sensing prey. In most groups of scorpions there are rows of small granules or tubercles in constant patterns on the pedipalps, carapace, and abdomen that are important in identification. Most scorpions are some shade of brown, from nearly black to pale sandy tan, often with two or three rows of darker spots on the body that may form longitudinal stripes. Bright colors or strongly contrasted patterns are uncommon in scorpions.

Almost every order of Arachnida is characterized by some type of unique sensory organ, and scorpions are no exception. On the ventral surface of the body, springing from between the bases of the last legs, is an inverted V-shaped organ, the pectines, with two arms. All scorpions have pec-

Shown in a characteristic scorpion defense and threat pose is *Paruroctonus mesaensis,* a common species from Arizona and New Mexico. Photo by Dr. Sherman A. Minton.

Most American scorpions belong to the families Buthidae and Vaejovidae and have relatively slender and weak pedipalps. Several species of *Diplocentrus* (family Diplocentridae), however, occur in the southwestern U.S. and Mexico and have very inflated pedipalps that give them a ferocious appearance. Photo by Dr. Sherman A. Minton.

Nebo, an Old World (eastern Mediterranean) diplocentrid scorpion with very impressive pedipalps. The presence of inflated pedipalps usually means the scorpion is harmless or nearly so.

Lychas marmoreus, a slender scorpion from New South Wales, Australia. Photo by Dr. Sherman A. Minton.

tines. Each arm (a pectine or pecten) carries a variable number of teeth, the number usually differing from species to species and occasionally from male to female within a species. Although some scorpions have only five or six teeth per pectine, most usually have 20 to 30 or more. No one is quite sure about the function of the pectines, but they seem to be sensory organs of some type, perhaps of touch, humidity, or even tilt of the substrate.

Sexes are difficult to distinguish in most scorpions, though there are usually minor differences in size of the pedipalps (broader in males), number of teeth on the pectines (more numerous in males), or strength of crests on the abdomen (stronger in males). In both sexes there is a rounded genital operculum just above the base of the pectines that covers the openings to the genital system; in male scorpions of some species genital papillae may project slightly beyond the operculum. When mating, males go through a sometimes elaborate courtship "dance" while facing the female and holding her pedipalps. Part of the "dance" may actually be an attempt to use the pectines to find the proper hard and flat substrate on which the male will deposit a complex spermatophore, a capsule-like structure that contains a mass of sperm. The male then guides the female over the spermatophore and she picks up the active part of it with her genital operculum. Fertilization is internal. All scorpions give birth to living young that are fully formed miniatures of the parents, gestation taking several months. Immediately after birth the young scorpions (from six to over 50 in number) move to the mother's back and are carried about by her until they have molted and used up their yolk. Interestingly, different families of scorpions vary in

Scorpions usually digest (at least in part) their food outside the body, mauling it with the chelicerae and leg bases and covering it with saliva before actually eating it.

the way the young are arranged on the mother's back, from very haphazard and piled at random to arranged in constant rows at regular intervals. Young scorpions can sting effectively as soon as they are born.

Of the five or six generally recognized families of scorpions, only the family Buthidae contains seriously venomous species. The very large scorpions such as species of *Pandinus* and *Heterometrus* from the Old World (family Scorpionidae) and the hairy *Hadrurus* of the southwestern U.S. and Mexico (family Vaejovidae) are virtually harmless, seldom inflicting a sting more painful than a beesting (although allergic reactions are known to occur on occasion). Of the 1500 or so scorpion species, only about 50 are considered a serious threat to human lives, and most of these belong to the genera *Buthus, Androctonus, Leiurus,* and *Buthotus* (Old World) or *Centruroides* and *Tityus* (New World).

OLD WORLD VENOMOUS SCORPIONS

The dozen or so species of *Androctonus* are very heavily built scorpions with large and often serrated keels on the abdominal segments. Like most of the other venomous Old World scorpions, they are usually about 4 inches long, inhabit dry areas, and have relatively small and slender pedipalps for their body build. (Most buthid scorpions, thus almost all dangerously venomous species, have slender pedipalps. Very broad pedipalps usually mean the scorpion is harmless.) In most of these scorpions the color varies from dark brown to sandy yellowish tan, often variable in different populations or individuals of a single species. The species of *Androctonus* are found from the drier regions of

Buthotus species, a venomous scorpion from the Sind Desert of southern Pakistan. Photo by Dr. Sherman A. Minton.

North Africa through the Middle East to India. Probably all the species are at least dangerous, but human fatalities from the stings of *A. australis* (North Africa) and *A. crassicauda* (North Africa and Middle East) are especially common. As in all other venomous scorpions, small children are more susceptible to death from scorpion stings, with fatalities up to 20% reported.

Buthus occitanus is one of the more common scorpions of the Mediterranean countries of Europe and North Africa, and it is also one of the more dangerous species. Although in southern Europe it is not considered dangerous to man, in North Africa it is a proven killer of children under the age of 10. Adults of this species seldom exceed 3 inches in length.

In India and Pakistan the most dangerous scorpion is considered to be *Buthotus tamulus*, a 3- to 4-inch species often found under stones in villages. Again, it is most dangerous to children.

Leiurus quinquestriatus is widely distributed from North Africa to Turkey and the Middle East. This 4-inch species is one of the most toxic scorpions as determined by laboratory tests on mice, but it usually produces less venom per sting than the *Androctonus* species and is thus somewhat less dangerous to adult humans.

NEW WORLD VENOMOUS SCORPIONS

The most familiar venomous scorpions in the Americas are members of the genus *Centruroides,* if only because this genus occurs in the United States. Of the two dozen or so

Centruroides sp. from Durango, Mexico. This very long-tailed species is similar in general shape and color to *C. exilicauda* from northern Mexico and the U.S. but probably represents one of the numerous other species of Mexican *Centruroides,* many of which are seriously venomous. Photo by Dr. Sherman A. Minton.

species, very few can be readily identified and the taxonomy of the genus is very uncertain. Most are small pale brown scorpions often with darker stripes on the body and tail; they are slender, with long legs and tails, but seldom exceed 3 to 4 inches in length. These scorpions are secretive nocturnal species that hide by day in cracks and crevices and under debris, including such artificial debris as wood piles, tents, sleeping bags, and clothing. They often enter houses in Mexico if the floor of the house is not several inches above ground level.

Because the taxonomy of the genus is so confused, it is difficult to mention individual species. Several dangerously venomous species occur in northern and central Mexico, but in the United States the only dangerously venomous species is *C. exilicauda*, which occurs in much of Arizona and in adjacent states as well as northern Mexico. *C. exilicauda* has killed several children and a few adults in Arizona. To show the extreme taxonomic confusion in the genus, until recently this species was called *C. sculpturatus* and *C. gertschi* in the United States, with *C. exilicauda* thought to be an obscure Baja California and northwestern Mexico species. *C. exilicauda* was described in 1863, *C. sculpturatus* in 1928, and *C. gertschi* in 1940. Thirty years after it was described, *C. gertschi* was recognized to be a color variety of *C. sculpturatus*, but it took another ten years before it was recognized that the Arizona *C. sculpturatus* was inseparable from the Mexican *C. exilicauda*.

Over almost all the southeastern United States the scorpion *C. vittatus* is the only species to be found (other species occur in Florida). Although it is common in many areas, it fortunately is not dangerously venomous, its sting seldom producing more than a sharp pain that disappears in three or four hours after some local swelling. This has not prevented local inhabitants from exaggerating its venomous nature, however, and the species is greatly feared throughout its wide range.

The genus *Centruroides* is rather unusual in that only a few of its species have actually been known to cause serious envenomation, with most species harmless but other morphologically almost identical species being deadly venomous.

The South American and Antillean species of the genus *Tityus* number about 40 or more and are all very similar. They are much like the species of *Centruroides* in general appearance but more commonly have keels and serrations on the abdominal segments of the tail. At least five of the species are considered dangerously venomous, with two Brazilian and a Trinidad species especially notorious. *T. serrulatus* of eastern Brazil is one of the most dangerous

An unidentified species of *Centruroides* showing the striped pattern typical of many species, including most of the non-venomous ones. Photo by Dr. Sherman A. Minton.

species of scorpion, being responsible for over 100 deaths each year, most of them of children. In addition to its toxicity, it has the unfortunate habit of living in and near houses, thus putting it in ready and constant contact with humans, including small children. This Brazilian species is most unusual in that it seems to be parthenogenic, as no males have been found in the over 100,000 specimens examined in venom laboratories. *T. bahiensis* from Argentina and southern Brazil and *T. trinitatis* from Trinidad and Venezuela are also very dangerous house-dwelling species that cause numerous deaths each year.

SCORPION STINGS

The stinging apparatuses of all scorpions are very similar. The venom glands are large and capable of introducing large amounts of venom into the victim; additionally, cornered or crushed scorpions tend to sting repeatedly. Usually the sting is instantly painful, though in *Centruroides exilicauda* the pain can subside so rapidly it may be forgotten immediately. There may or may not be local swelling, but it is seldom severe in the dangerously venomous species. In *C. exilicauda* swelling is absent, but there is an increased sensitivity to pain, so the slightest touch to the skin near the sting is greatly magnified and may cause screaming or convulsions. The venom in scorpions is strongly neurotoxic, and death results from respiratory and cardiac failure. Minor and inconstant symptoms that vary from species to species include such things as increased perspiration, blindness or double vision, involuntary muscular spasms, difficulty in speaking, and a swollen tongue. To adults, a scorpion's sting is seldom fatal even from the proven deadly species, but children are extremely susceptible to virtually all the dangerous species mentioned. The smaller the child, the more serious the reaction. Although in adults untreated mortality rates from stings of most species are under 1%, in school-age children they may rise to 5-10%, and they often exceed 20% in babies.

Treatment is usually supportive. Antivenin is produced for the stings of several species but is hard to obtain except in the areas where scorpions are the greatest problem, such as North Africa, eastern Brazil, or Arizona. In the case of bites by *Centruroides* species, immediately placing the hand or foot in ice water helps slow the absorption of the venom until antivenin can be obtained. The immersion must be done within five minutes to be effective; chunks of ice and medicinal spray coolants also work. This simple therapy is also supposed to be very effective in reducing pain and swelling from stings of nondangerous species.

In South America the species of *Tityus* take the place of *Centruroides* as the most prominent venomous scorpions. Like most venomous *Centruroides,* the dangerous species of *Tityus* are slender and lightly built.

8: Spiders

Spiders are the most commonly seen arachnids and also the most repulsive to many people. Just why people are afraid of or repulsed by spiders has been discussed in many scientific and not so scientific papers, but there is general agreement that much of it has to do with their common habit of lurking in the shadows and corners, their strange feeding habits (at least strange to vertebrates such as man), and their usually soft, rounded abdomens. Additionally, it can be truthfully said that with only a handful of exceptions, all of the 50,000 or so described spiders are venomous animals, as all possess poison glands in or at the bases of their mouthparts. Certainly the notoriety gained by spiders such as the black widows and brown recluses has not helped the spiders to be more tolerated by the common man.

Though spiders vary in size from less than 1/50 of an inch to over 4 inches in body length (with their long legs always making them seem larger than they really are), they are fairly uniform in structure. All have four pairs of legs that are generally very similar in size and shape, most have six or eight pairs of eyes (though many spiders have fewer eyes or lack them altogether), and all have chelicerae (mouthparts) that end in large, curved dark fangs. Anterior to the first pair of walking legs is a pair of pedipalps that are very leg-like in shape and often in size and are modified in males to transfer sperm to the female. The head and thorax are covered by a carapace and are distinctly separated from the usually globose and soft abdomen by a narrow waist. The abdomen may be elongated in some species and may even have decorations such as spines in a few, but there is no visible segmentation except in a few primitive groups. At the end of the abdomen are the spinnerets that produce the silk used in webs, egg capsules, and the linings of burrows.

Although few species are dangerously venomous, tarantulas are feared by most people. Shown above and on the facing page are two large theraphosids or bird spiders from South America. It is practically impossible to identify tarantulas even with specimens in hand, so any names used here should be considered very tentative and possibly incorrect. The Brazilian and Argentine species above has been called *Grammostola actaeon* by the photographer, Pedro Antonio Federsoni.

As mentioned, almost all spiders (except two small and obscure families) have venom glands. These glands are usually rather cylindrical and elongated and are covered with heavy coats of muscle. A duct leads from each gland to the fang of the chelicera. Since the coats of muscle seem to be under voluntary control, the spider can determine if a bite will also inject venom into the prey or enemy and can determine whether a large or small dose of venom will be released. This voluntary aspect of venom release has caused much confusion when laboratory tests have been run to determine if a particular species of spider is actually seriously venomous.

In most spiders the males are smaller than the females and often have slightly different colors or patterns and very often have unusual carapace shapes. Additionally, the male pedipalps have modifications for sperm transfer. In the tarantulas and related spiders the last segments of the pedipalps are only slightly modified, but in the more advanced spiders the last segments are strongly modified, with many special structures. Sperm is released as small droplets from the anterior ventral part of the male's abdomen and scooped up by the specialized pedipalps. It is

The most commonly seen large tarantula of the south-central U.S. is *Dugesiella hentzi,* a 3-inch species often found wandering on the highways during mating season. This specimen was photographed in White Co., Arkansas, near the eastern edge of the range, by Dr. Sherman A. Minton.

Aphonopelma species are commonly sold as terrarium animals and are usually considered harmless (there appear to be no recorded ill effects from their bites). However, laboratory tests of their venom show them to be potentially dangerous. Photo by Ken Lucas at Steinhart Aquarium.

then transferred to abdominal sperm receptacles on the female, usually after a courtship dance or ritual of some type that serves to distract the always-hungry female from considering the male as a morsel of food. Even in non-active males the pedipalps are usually darker than those of the female and are often conspicuously bulb-like, making males easy to recognize. There is little truth to the old saw about female spiders always eating their mates—although it does happen on occasion, the courtship ritual usually serves to keep the female busy during and immediately after mating, at least long enough for the male to leave in peace.

Spiders are all carnivorous, usually feeding on small insects of appropriate size, although there are records of some larger spiders actually capturing and eating small birds and lizards. The prey is either trapped in a web or pounced upon (many spiders build no webs), subdued by the venom, and usually wrapped in a silken shroud that prevents it from struggling. Like the scorpions, spiders have external digestion, pouring digestive fluids on the food while mangling it with the bases of the mouthparts and sometimes of the front legs to aid digestion; when the prey is sufficiently liquid, it is transferred to the mouth. Some spiders actually suck the juices and semiliquid flesh from the insect food, leaving behind a perfect but empty chitinous shell.

Although there are probably more species of spiders in tropical countries, spiders can be found almost everywhere and often in great abundance both as to number of species and number of individuals. Even in the northern United States it is readily possible to collect 50 or more species almost anywhere without special effort, and even as small a state as Connecticut (which admittedly has been very well studied) has over 600 species recorded from its boundaries.

The normal tendency of most people is to think of any large and hairy spider (especially tarantulas) as being deadly venomous, this idea being strongly influenced by bad movie and television plots. Actually, the most venomous species in any locality (certainly in cooler climates) are likely to be small and rather featureless spiders such as can be found hiding in a cobweb in the corner of the basement or an abandoned shed. In the tropics there are admittedly several large hairy spiders that have been proved to be dangerously venomous, but these are usually local pests only and will be mentioned just in passing. We'll divide the dangerous spiders into four groups for our discussion: tarantulas (few of which are even very dangerous, let alone deadly), widows (the most common deadly species), brown recluses (deadly and now becoming more publicized), and miscellaneous dangerous types.

TARANTULAS

Although the large hairy spiders known as tarantulas or bird spiders are widely feared wherever they occur, few species are actually dangerous to humans. Of the dozen or more families recognized as tarantulas by systematists, only two (Barychelidae from Africa and some Australian and South American Dipluridae) contain species that have been shown to actually be dangerous, though it is suspected that a few Caribbean tarantulas also may be dangerous. The common American tarantulas belong to the family Theraphosidae and are relatively harmless. It must be mentioned, however, that laboratory tests of *Aphonopelma* venom show it to be about as potent as the venom of the scorpion *Centruroides exilicauda*—it is thus best not to be bitten by any tarantula, just in case.

Tarantulas are primitive spiders belonging to the suborder Orthognatha (also called Mygalomorpha, thus the common name mygalomorphs for all tarantulas) and differing in several important respects from all other living spiders. The most obvious difference is in the placement and movement of the fangs of the chelicerae. In most other spiders the fangs are angled in toward the midline of the body and thus move from the middle out when opened (perpendicular to the axis of the body). In tarantulas the fangs are parallel to the midline of the body or nearly so and thus open up and down. Most tarantulas are large spiders,

Few tarantulas are very colorful. This unidentified tarantula appears to be a male, as the modified pedipalps and hooks under the first legs can be made out in the photo .

Baboon spiders include some species (of the genus *Harpactirella*) that are supposed to be dangerous to man. Although they are usually harmless they are still very aggressive. Their classification is based on the rows of strong bristles under the legs that are used in digging burrows—as you might expect, they are not easy to identify.

with many species over 1 inch in body length and several exceeding 3 inches, and most are covered with brittle hairs on the abdomen and legs. In male tarantulas the pedipalps are only slightly modified for sperm transfer and are never as complicated as in most other spiders. Like other spiders, all tarantulas have venom glands, usually within the bases of the chelicerae.

Although North American tarantulas are uniformly non-dangerous, there are dangerous tarantulas in other countries. The African baboon spiders (family Barychelidae, genus *Harpactirella*) are large, long-legged hairy tarantulas that when disturbed tend to be very aggressive, raising the front legs and even attacking. Although their bite is usually no more offensive than a beesting, there are records of more serious effects, perhaps because of individual allergic reactions.

The other truly dangerous tarantulas belong to the family Dipluridae, a group of rather small (less than 2 inches) and smooth tarantulas with long spinnerets. These little tarantulas build large sheet webs of dense silk leading to a tube to one side in which the spider lives. When an insect becomes trapped in the web, the spider attacks it from its tube. In Australia the sheet-web tarantula *Atrax* is dangerously venomous and has caused human fatalities. The South American *Trechona venosa* is also considered to be dangerous to humans, but there seems to be some confu-

Atrax sp., a small black sheet-web tarantula from New South Wales, Australia, that is known to be dangerously venomous. Bites are common enough and severe enough that an antivenin has had to be developed. Photo by Dr. Sherman A. Minton.

Shown above is the large Brazilian *Theraphosa blondi,* one of the largest tarantulas. Notice that compared to the unidentified black tarantula shown below the abdomen appears almost bare posteriorly. This bald patch is due to the rubbing off of the special barbed hairs from the abdomen to "throw" at attackers; the tarantula below is assumedly younger or at least more freshly shed and has not yet had to use as many hairs. Photo above by Pedro Antonio Federsoni; that below by Ron Reagan.

Two large South American tarantulas.
Above: *Grammostola burzaquensis*, one species of a very large genus found over much of South America.
Below: A specimen of *Acanthoscurria;* some species of this genus are reputed to be dangerous to man.
Photos by Alcide Perucca.

sion as to whether or not it is truly responsible for the bites that have been attributed to it. In the United States and Europe, diplurids are uncommon, small, and not considered to be dangerous.

The very large tarantulas of the family Theraphosidae are the ones usually seen in movies and on television. They are also the most commonly encountered species in the deserts of North America and the jungles of South America. The species of *Grammostola* from South America may exceed 8 inches in diameter including the legs, but the more normal size in American theraphosids is less than 4 inches in diameter, with body lengths of 2 to 3 inches typical. In many of these tarantulas the males wander about in search of females during certain seasons of the year and are thus commonly found crossing roads or on lawns. Males are usually smaller than females and have hooks on the legs to help hold the female during mating. Although South American species of *Acanthoscurria* have been accused of being dangerous to man, there is little evidence to support this view. Also, some poorly known Caribbean tarantulas are reputed to be dangerous, but again evidence is lacking. The bites of theraphosid tarantulas are generally harmless or nearly so, though they may be very painful. Laboratory tests indicate that caution should be displayed, however.

American theraphosids have another defensive adaptation in addition to their fangs. This is the presence of special barbed hairs on the abdomen of most species. (Curiously, Old World theraphosids lack these special hairs but have much more aggressive threat behaviors.) When a tarantula is cornered, it brushes its hind legs over the abdomen, releasing a cloud of fine hairs. These are irritants but are not venomous, causing local skin rashes in humans and perhaps mild respiratory distress if too many enter the mouth or nose. Several major types of hairs have been identified, and it also seems that each species of tarantula might have hairs that are slightly different from those of the next species. The hairs of some tropical genera are capable of killing or disabling mice, presumably the natural predators of these tarantulas. Although the tarantulas commonly kept as pets (mostly species of *Dugesiella, Aphonopelma,* and *Brachypelma*) are not particularly noxious, it is still best not to handle them any more than necessary or to breathe the air in their containers.

WIDOWS

Black widow spiders (genus *Latrodectus*) are common inhabitants of most of the United States. According to the literature, the most common species, *L. mactans,* occurs

Widows usually hang head-down from their irregular webs, and females seem to almost always be guarding an egg case. The complete red hourglass on the abdomen of this female means the species is probably *Latrodectus mactans.*

over most of the world, from Canada to the southern tip of South America, in most of Africa, in southern Europe, in the warmer parts of Asia, and in Australia. However, there has been a recent tendency to doubt this distribution as it is now known that the taxonomy of the black widows is not as simple as once thought, with several and perhaps many species and subspecies being involved in this tremendous range. For this reason it becomes difficult to describe a typical widow or to list all the names involved.

Most species of *Latrodectus* are about ½ to ¾ inch long, with long legs and a round abdomen. The body usually is black or blackish brown, and there are usually at least traces of a red to orange hourglass or double triangle marking under the abdomen. There may be other red or brown spots or stripes on the abdomen as well, and occasionally white spots (one species has a white abdomen). In the United States the common species are the eastern *L. mactans,* often with a row of red spots down the middle of the abdomen and a complete hourglass below; the western *L. hesperus,* with the abdomen unmarked above; and the northern *L. variolus,* with white bands on the sides of the abdomen and the hourglass broken into separate triangles. These species are also widely distributed on the other continents as well, although the details of their distribution are just being worked out. All seem to be equally venomous, however.

The widows belong to the very large family Theridiidae, the comb-footed spiders. The members of this family build irregular webs (cob webs), often in corners of houses; the most common house spiders are members of the Theridiidae. These spiders are called comb-footed spiders because of the presence of a row of short comb-like bristles on the back legs. They have the habit of disabling prey trapped in their webs by throwing great quantities of almost liquid silk over the struggling food, the hindlegs with their combs doing the throwing. Although some of the common house spiders of this family are suspected of being at least potentially dangerous, only the widows are proven harmful.

Widows build their snares in quiet corners of outbuildings, near garbage dumps, under stones, on the sides of houses, in bushes, and almost anywhere else that is dark and secluded. Only a single female usually is found on a web. At rest she stands upside-down, often guarding a rounded brownish egg case containing numerous eggs. The female widow is extremely timid and retreats to the furthest corner of the web and freezes if she is disturbed, sometimes not appearing again for hours. Widows are not aggressive and do not attack even if provoked—bites only occur through accidentally crushing a spider against the skin or through deliberately mishandling one. The short-lived males (which are seldom eaten by the females) are about a quarter to half the size of an adult female, are brightly colored, and are so small that they cannot bite and are thus harmless.

For many years scientists doubted that widows were actually dangerously venomous to humans. Although native peoples respect widows wherever they are found and usually have common names for them (a most unusual circumstance for a small spider), their beliefs that the spiders were dangeorus were often considered to be folk tales. Even as late as 1936, researchers found it necessary to ask the question, "Does the spider represent a real menace?" and answer it by saying that "The statement sometimes heard, even from zoologists who should know better, that the bite of a black widow is no more dangerous than that of a mosquito, must, in view of extensive clinical experience, be branded as false and dangerous" (D'Amour, et al., *Quart. Rev. Biology*, 11(2), June, 1936). There is no doubt that widows are deadly at times and that there are many human fatalities that can be attributed to bites from the various species. Although the fatality rate in untreated bites is probably less than 5%, one or two people die each year in the United States from their bites.

The problem is that the effects of widows' bites are totally unpredictable. Because the venom glands are under voluntary control, it is possible to be bitten and show no symptoms at all because no venom was injected. In other instances only a small amount of venom is injected, with

Americans usually think of *Latrodectus* species as "black" widows, but in other parts of the world they may be "brown" widows or even, as in the case of the *L. pallidus* shown here, "white" widows. Photo by Dr. Sherman A. Minton of a specimen from Israel.

minimal symptoms, but the same spider can later produce a full bite causing severe symptoms or even death. Since widows are often domestic spiders and may be very abundant in almost any area of the world, their chances of accidental contact with humans are increased, making them an especially serious hazard. The shiny red and black pattern plus the freezing behavior of the female also draw the attention of children, who may be tempted to pick up a spider as a toy.

In humans, if symptoms develop they follow a very constant pattern. There is little or no pain at the time of the bite, but from 10 to 60 minutes later a dull pain begins to spread from the site of the bite up the limbs and into the back, chest, and abdomen. The pain soon becomes extremely severe and is difficult to control even with pain killers (though hot baths help). The abdominal wall becomes rigid and board-like, resembling the condition found in a ruptured appendix. Frequently present are nausea, cramps, vomiting, increased blood pressure, and increased salivation and perspiration. The pain may be almost continuous or only in fits, and there may be the unusual symptom of extreme pain even in the soles of the feet.

In most cases these symptoms continue for only one or two days, and recovery is certain though slow. If antivenin is administered, the symptoms subside quickly and the patient may not even have to be hospitalized. Death is very infrequent and is most common in children who have suffered a full bite. Even full bites on healthy adults are very rarely fatal or even serious, but the potential for death or major injury from the bite of any widow should never be underestimated.

BROWN RECLUSES OR FIDDLEHEADS

The long-legged brownish species of the genus *Loxosceles* are common in many areas of the southwestern United States and range into southern South America, preferring dry, littered habitats with plenty of retreats. Only recently has it been determined that the bite of brown recluses is dangerously serious to man. Although for many years the South American *L. laeta* and similar species have been considered at least theoretically dangerous, only in the last 20 years or so have the North American species been determined to be almost as dangerous.

Brown recluses belong to their own family, Loxoscelidae, and are close relatives of the spitting spiders (Scytodidae) that are common house pests. Like their relatives, *Loxosceles* species are domestic spiders when the occasion arises, preferring dry, dark corners or closets with litter such as paper or old clothes. Brown recluses are seldom

A living specimen of *Loxosceles laeta* from South America. This very dangerous species is a serious medical threat in parts of South America and has been introduced (but not established) in several American ports. If you look closely you can see a faint "violin mark" on the cephalothorax and that the eyes are in three groups arranged in a semicircle rather than just a cluster as in most spiders. Photo by Alcide Perucca.

This close-up view of a dead brown recluse spider, *Loxosceles reclusa,* shows the slenderness of the legs, the definite "violin mark," and the three pairs of eyes. The complex pedipalps show this to be a male. Photo by Dr. Sherman A. Minton.

seen during the daytime, doing much of their hunting at night, when they may even venture onto beds. The species are all very similar, being various shades of brown with long brownish legs. There is usually a darker brown marking on the front middle of the carapace that looks a bit like a violin with the neck pointing backward, thus the common name "fiddlehead spiders" for the group. When viewed with a hand lens, it can be seen that they have only six eyes, not the usual eight, and the eyes are in three pairs in a widely spaced semicircle across the front of the cephalothorax. Brown recluses can be confused easily with another house spider, *Filistata hibernalis,* that is similar in color and may even have the violin-like marking, but in that species the legs are shorter and heavier and there are eight eyes in a closely spaced cluster toward the middle front of the carapace. *Filistata hibernalis* has been blamed for a few mild cases of human envenomation but is usually not considered to be venomous.

The venom of *Loxosceles* is almost exclusively necrotic in action, destroying muscles and fat. At the time of the bite there is a mild burning or stinging pain, but nothing serious. In two to eight hours the site of the bite becomes red, the pain increases, and there may be blister formation around the discolored area. Later the middle of the wound forms a firm star-shaped deep purple area, and soon the skin and muscle cells around the wound start to slough. This process of necrosis and sloughing of dead tissue may continue for two or three weeks, with formation of a large, deep scar. Death occurs in about 6% of untreated cases if complications do not occur, so the species are definitely dangerous.

Serious complications arise in a small but significant percentage of brown recluse bite cases. In these cases, called the viscerocutaneous form, the internal organs, especially the kidneys, are affected by the venom and become necrotic. There is sudden and severe anemia, with internal bleeding and fever. Although this form occurs in only 13% of bite victims, almost 30% of the cases of viscerocutaneous involvement are fatal, an extremely high mortality rate for invertebrate venom. As usual, all symptoms are most severe in children.

Prompt medical treatment is necessary, as most of the effects occur or begin within the first day after the bite, and deaths usually occur within one to three days. Excision of the necrotic tissues at the site of the bite followed by skin grafts is the usual treatment; there is no real treatment for the viscerocutaneous form. Antivenin is seldom available in the United States.

As long as brown recluses were outdoor spiders hiding under rocks in the deserts and plains of southwestern North America and southwestern South America, they were really not problems to anyone. The spiders are secretive, seldom venture far from cover, are totally nonaggressive, and bite only if squeezed between bare flesh and clothing. However, these spiders are easily carried about with man and have now been introduced, apparently successfully, over much of the southern and southeastern United States at least as far north as New Jersey and Massachusetts. They readily become established in houses and sheds, bringing them into close contact with humans and thus allowing more accidental bitings to happen. Even the South American *L. laeta*, a species that is much more dangerous than the North American species, has been introduced into several U.S. ports and readily begins to reproduce, surviving attempts at eradication. In all likelihood, brown recluses will continue to be an important source of spider poisoning in the United States and the number of incidents may grow significantly as the spiders become more common in populated areas.

MISCELLANEOUS SPIDERS

As taxonomic and medical research continues on spiders and the reportedly harmful effects of their bites, more and more species have been found to be at least potentially dangerous to humans, though often in only minor ways and perhaps only as uncommon allergic reactions. Thus the list of suspected dangerous species grows every year, but few

Any large spider that comes into ports with tropical produce is likely to be called a "banana spider." Unfortunately, some of these hairy spiders actually are at least potentially venomous and may cause serious envenomations in workers unloading produce. This is an unidentified huntsman spider, possibly *Phoneutria*, from South America.

species are actually proved to be dangerous enough to humans to be of medical concern. There are at least three exceptions, however, that should be briefly discussed, even though two of them will probably never be of importance to most readers of this book.

Among the most impressive of the true spiders (*i.e.*, non-tarantulas) are the South American huntsman spiders, *Phoneutria* (family Ctenidae), also known as wandering spiders or banana spiders (actually, the name banana spider is used for any large, hairy spider that travels into foreign ports in bunches of bananas). They are nocturnal hunters that build no webs and hide during the day. They are among the largest non-tarantula spiders in South America, with body lengths often exceeding 1 to 1½ inches and leg spans of over 5 inches. The legs are hairy, as is the entire body, and there often are complicated color patterns on the chelicerae and abdomen that are presumably used in mating displays. When cornered the spider drops its abdomen and back legs to the ground, raises the carapace to display the often brightly colored chelicerae, and raises the two front pairs of legs straight up into the air, producing a very menacing effect. Actually, these spiders are somewhat aggressive and may bite on small provocation. In winter they tend to move into houses for warmth in southern Brazil, placing them in close contact with humans and resulting in many bites.

This large Brazilian *Phoneutria* species displays the bright colors typical of the group. The yellow leg bases and reddish brown chelicerae of this species are presumably important in mating and threat displays. These are aggressive spiders and are important sources of human envenomation in South America. Photo by Dr. Sherman A. Minton.

The effects of a bite from a huntsman spider are intense radiating pain, cold shudders, an irregular pulse, cramps, possibly convulsions, increased perspiration and salivation, vertigo, and tightness of the chest. The venom is mildly neurotoxic, and there is little more than redness to mark the site of the bite. From most bites recovery is complete (except for sensitivity around the site of the bite) within 12 hours, but full bites in small children or weak adults may cause death within the first five hours after the bite. Antivenin is produced in Brazil, where the spider is a major pest.

Like other large South American spiders, these often stow away on bunches of bananas and other produce and may be introduced into foreign ports, including those in the United States and Europe. Although they cannot adapt to the cooler climates, they may be active for several days after introduction. There are a few cases reported of bites on cargo handlers and others who accidentally came into contact with them in the U.S. and Europe.

In northern climates the wolf spiders, family Lycosidae, are familiar and common large spiders (often half an inch or more in length, with long legs to increase their apparent size) that may enter houses but are considered harmless. They are often striped with dark brown on gray or pale brown. Since they are solitary hunters, they appear rather intelligent for spiders and are sometimes kept as pets. Their requirements are simple—a few flies and access to water. Females carry the large round egg sac attached to the back of the abdomen, and after the young are hatched they ride about on the mother's back for a few days before going their own way. One wolf spider, *Lycosa tarentula,* was the species blamed for outbreaks of supposed spider poisoning in southern Europe that could only be cured by dancing the tarantela.

In South America, however, the species *Lycosa raptoria* and *L. pampeana* are considered to be dangerously venomous. Like that of brown recluses, the venom is necrotic in action; they usually inject large quantities of venom when biting, increasing the severity of the bite. The venom does not cause a severe local reaction, but about 20 hours later the site of the bite turns ashy gray surrounded by reddish discoloration. Usually the pain at the site remains dull and is seldom incapacitating. A few days later the center of the wound darkens and tissue begins to slough, so when the scab is shed a deep ulcerated pit is formed and keeps growing through tissue necrosis for several more days or even weeks. Scars up to 8 inches long have been reported. There are no internal complications as in bites of *Loxosceles,* and death apparently seldom or never results.

Polybetes pythagoricus, a South American huntsman spider of the family Sparassidae. Photo by Rogelio Gutierrez.

A fairly typical wolf spider, the South American *Lycosa poliostoma.* Female wolf spiders often carry egg cases with them as shown here. Photo by Alcide Perucca.

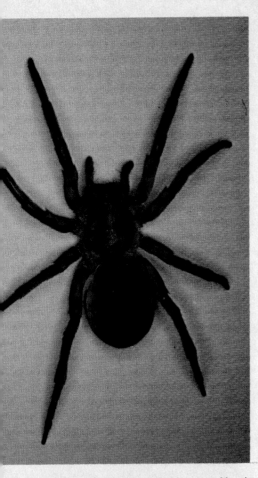

Lycosa carolinensis, a common North American wolf spider. North American members of the Lycosidae are considered harmless and may even be kept as pets, while those from South America have bad reputations. Photo by Dr. Sherman A. Minton.

The last spiders we will discuss are the small domestic running spiders of the genus *Chiracanthium* (family Clubionidae) that are common in shrubs and under stones in gardens. They are usually about ½ inch or so long and have stout round abdomens and pale bodies that may be greenish to yellowish or whitish without strong color patterns. Often they are seen running about after small insect prey near the tops of low shrubs or near the ground, where they have white silk cocoons used as retreats, winter hibernation sites, and nests. Two species are common in the eastern United States, with several more species known from the rest of the world, especially in cooler climates. *Chiracanthium inclusum* is native to the eastern United States and is most likely to be found in gardens, while *C. mildei* is introduced from Europe and is more likely to be found actually inside houses.

Only recently have these spiders been implicated in human envenomations, and little is yet known about the severity or occurrence of their bites. They tend to defend their cocoons from gardeners or housewives who accidentally brush across them, resulting in bites that produce local swelling, redness, and fever. A sharp, radiating pain is also usually present, and there may be sloughing of tissue from around the site of the bite as in brown recluse bites. No internal complications are known, and the bites are usually mild with no severe symptoms or scarring. Allergic individuals may have more severe reactions, however.

As concerns spiders and almost any other group of animals discussed so far, it is best to consider all of them as potentially venomous and avoid them where possible. Don't handle spiders, don't slap at them with the hand or step on them with bare feet, and try to prevent formation of suitable domestic habitats for spiders by not letting litter build up in closets, basements, and outdoor sheds.

One last aspect of spider poisoning should be mentioned. This is the growing tendency to blame almost any type of mysterious illness involving tissue destruction (necrosis) on bites of brown recluses. Knowledgeable medical practitioners make it a point to be sure that a spider bite is actually involved in an illness before blaming a spider. Brown recluses have been blamed in print for bites that occurred in parts of North America from which they are not recorded and have been blamed for illnesses in which tissue necrosis was only a minor aspect of the problem. Brown recluses and other spiders are usually shy and not aggressive, and they are not likely to cause epidemics of bites in an area. Blaming spiders for bites without hard evidence in the form of a specimen is not good medical practice.

9: Dangerous Insects

At least 500,000 species of insects are known, with the actual number of insects in existence certainly exceeding 1,000,000 and possibly several times this amount. Most insects are beetles (which are seldom dangerous although a few blister beetles of the family Meloidae and a few species of other families release a burning fluid when disturbed), moths and butterflies (of which the caterpillars are sometimes venomous), true flies (important as carriers of diseases but seldom venomous except for local reactions to saliva of mosquitoes and deerflies), and the hymenopterans or wasps, bees, and ants (where most of the venomous species occur). In the United States and European countries, most fatalities due to bites or stings of venomous animals can be traced to stings from bees and wasps, with 229 of 460 total venom fatalities in the U.S. from 1950 to 1959 being due to hymenopterans, almost double the fatality rate due to snakebite. In tropical countries the death rate due to insect stings is also very high compared to other causes, although here the rate due to snakebites and other animal venoms increases compared to cooler countries. If you are to be seriously injured by any animal bite or sting, the chances are greatest that the cause will be a wasp, bee, or ant, as these animals are abundant everywhere, many species are venomous to some degree, and there is a strong tendency to multiple stings that may cause severe reactions.

Although there are records of local reactions to many different types of insect bites and stings (including such things as mosquito bites, flea bites, bedbug bites, or sensitivity to the scales of butterfly and moth wings) and there are even occasional deaths from such usually innocuous encounters with insects, most venomous insect encounters fall into three categories: severe skin rashes from stinging caterpillars; bites from true bugs; and stings from wasps, bees, and ants.

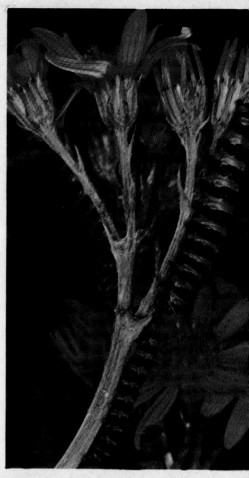

Many adult and larval insects are "protected" from predation by alkaloids or other noxious chemicals that are incorporated into the body fluids and tissues, the chemicals often being acquired from the plants eaten during larval stages. These brilliantly colored cinnabar moth (*Callimorpha*) caterpillars are typical of such noxious, but not dangerous, insects.

Facing page: Although the large, toothed jaws of this Australian bulldog ant (*Myrmecia*) would be capable of causing considerable pain, the stinger at the tail end is the part to be truly feared. Contrary to popular belief, it is the sting of ants that is venomous, not their bite.

STINGING CATERPILLARS

Caterpillars of butterflies and moths are usually innocuous small animals that feed on various types of plants. They make up the major portion of the diet of many types of birds and predaceous insects and are often "protected" in some way, either by noxious chemicals within the body (often obtained from their food plants), by behavior (feeding at night or various sudden displays of brightly colored patterns to scare off predators), or by sharp spines. In several unrelated types of moths and even a few butterflies (such as the South American morpho species), these defensive spines are more than simple irritants like the hairs on a tarantula. In these truly venomous caterpillars the spines are hollow, often barbed at the tip or glass-like, and connected at the base to large venom glands. If a hand or arm accidentally brushes across such a species, the spines easily penetrate the skin, break off, and allow the venom to enter the wound.

Although such an encounter is not fatal, it can be extremely painful. There is an immediate burning pain where contact was made, followed by numbness and sometimes by swelling. Often a tiretrack-like red scar marks the position of the caterpillar's spines where they touched the skin. There may be headaches, nausea, and even shock or convulsions in especially sensitive victims. Fortunately, these symptoms are usually short-lived, disappearing in just minutes or hours under normal circumstances. The spines can be removed from the wound with pieces of sticky tape, and topical pain relief products can then be applied. Medical attention is usually necessary only in extreme reactions.

Although almost any spiny caterpillar can cause these reactions, the most familiar culprits are various species of the moth families Eucleidae, Megalopygidae, and Saturniidae, at least in North America. Few saturniid caterpillars are harmful, but the caterpillar of the io moth, *Automeris io*, is a common source of accidents in the southern U.S. This chubby 2- to 3-inch bright green caterpillar has a thin red stripe and a white stripe along each side of the body. The stinging spines are in large tufts on the top and sides of each segment. This is a relatively harmless species, the pain being usually not nearly as intense or long-lasting as that of the following caterpillars.

The most spectacular of the North American stinging caterpillars is probably the saddleback, *Sibine* (Eucleidae). These inch-long green caterpillars have a brown saddle-like round spot in the middle of the back and usually some contrasting white spots or stripes outlining the saddle on the green back. At both ends of the body are long tubercles

Black and red are warning colors in much of the animal kingdom, and it is always wisest to be careful when handling any animal presenting such a pattern. This caterpillar is another protected species that is distasteful to predators but lacks secretions harmful to humans.

The prominent spines of this saddleback type of caterpillar are a good indication of its venomous nature. Although few people would voluntarily handle such a spiny animal, it takes only a casual contact with the venomous spines to produce a painful sting and sometimes more serious effects.

bearing venomous spines, plus smaller spine-bearing tubercles along each side of the body. The pain from a sting by these spines is intense and often long-lasting, sometimes with all the additional symptoms (nausea, swelling, etc.) previously mentioned.

Perhaps the most severe stings result from encounters with the small brownish caterpillars known as flannel moth caterpillars or puss caterpillars (Megalopygidae). The species of *Megalopyge* and related genera are literally covered with tufts of stinging hairs, often so densely that the back cannot be seen through the hairs. When these hairs come into contact with exposed skin, they produce a severe and exceedingly painful rash that may last for days. Often they also produce reddish blistered areas that look like tiretracks and may be visible for weeks.

TRUE BUGS

The order Hemiptera contains several thousand species of true bugs (as opposed to the use of the word "bug" to indicate any insect or vaguely insect-like animal). Although

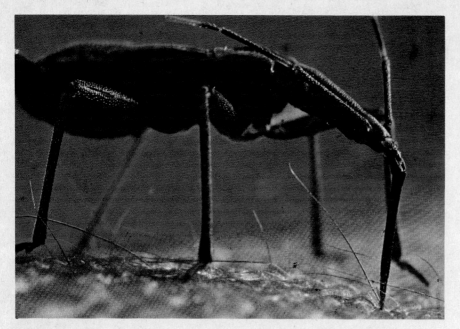

The prominent beak of an assassin bug is able to easily penetrate human skin; in fact, it has actually evolved to penetrate insect joint tissue, a much tougher substance than normal skin. Although the bites of assassin bugs may be painful, they are to be feared more for the parasites that their bite might transmit.

many are feeders on plant sap or various plant parts, others are predaceous, feeding mostly on other insects. The mouthparts are formed into a long or short beak that pierces the plant's or animal's skin to suck out body fluids. Often a special saliva is injected into insect prey that causes paralysis and also liquifies tissue near the site of the bite so it can be sucked by the bug. The most notorious true bugs are probably the bedbugs that feed on man at night in many parts of the world where standards of hygiene are not the best. Many other bugs are familiar to gardeners, such as the squash bugs and chinch bugs that feed on garden plants and lawn grasses.

Several commonly encountered families of true bugs have species that cause severe pain and sometimes other complications when they bite man. Some families of water bugs, including especially the backswimmers (Notonectidae), waterscorpions (Nepidae), and giant water bugs (Belostomatidae), have a potent saliva that causes pain like a beesting and sometimes local swelling. The backswimmers are usually small insects (less than half an inch long) that produce a bite that is much more painful than one would expect from their small size. Waterscorpions and giant water bugs are often 2 to 4 inches in length (including long siphon tubes) and their bites are often very painful. Some species of giant water bugs are attracted to light during mating season (thus the name electric light bugs) and may litter sidewalks under street lamps to a depth of several inches.

The assassin bugs, family Reduviidae, are the most dangerous of the true bugs. The couple of thousand species of this family are found throughout the world, with the greatest variety in the tropics. Most species are about ½ to

The "kissing bug" of the eastern United States, *Arilus cristatus.* This large grayish bug is readily recognized because of the cog-wheel-like spines on the prothorax. Although its bite is very painful and may cause minor side effects, its evil reputation is due mostly to exaggerated stories. Photo by Dr. Sherman A. Minton.

Most reduviids are black or gray in color like the species shown here, but some are stunningly patterned in red and orange.

1 inch long, with a few exceeding 1½ inches. Black is their predominant color, often offset with bright red or orange tones on the edges of the abdomen that project beyond the wings at rest. The legs and antennae are long, as is the beak, which curves back under the head. There are often spines on the body that are important in species identification.

Assassin bugs are active predators on other insects and occasionally on vertebrates. In the tropics several species have become adapted to human habitations and act much like bedbugs, coming out of cracks at night to feed on human blood. These blood-feeders often carry protozoan diseases such as Chagas disease, a trypanosome, making the aftereffects of their bite much worse than the bite itself. In non-tropical areas reduviids are usually encountered while gardening or working with plants or even near house lights at night. Usually their bite is painful for only a second or no pain at all is felt, but a reddish area soon develops around the site and there may be small blisters as well. The red area grows and turns brown. Eventually small brownish red nodules develop near the site of the bite and often on the extremities. All this is caused by the saliva injected into the wound by the bug, as it not only causes temporary muscular and nerve paralysis (on a human this effect is almost unnoticeable) but also causes tissue destruction and liquefaction. The symptoms usually disappear without treatment in a few days, although rare allergic reactions may complicate the condition.

The most commonly encountered assassin bug in the United States is the wheel bug, *Arilus cristatus,* a grayish brown to blackish bug about 1 to 1½ inches long that is common in the warmer states and is occasionally common further north. It is readily recognized by its size combined with the presence of a strongly curved ridge on the middle of the prothorax behind the head, this ridge bearing a row of short blunt knobs that give the whole thing the appearance of a cogwheel. Unlike most other assassin bugs, the bite of this species is extremely painful and may cause discomfort for up to five or six hours.

All the assassin bugs, including especially the wheel bug, are commonly called "kissing bugs" by the media based on an incident that happened in Washington, D.C., in 1899. A woman was bitten on the lip by a reduviid and the incident was blown up by the newspapers, resulting in an alleged outbreak of "kissing bug disease" throughout the nation—at least among readers of newspapers.

ANTS

Ants are social insects very closely related to the wasps and bees but distinguishable by various technical features such as the structure of the antennae and the presence of humps on the petiole connecting the abdomen and the thorax. In point of fact, ants are most readily recognized by making nests occupied by three castes: one or a few queens, many workers (sterile females), and (seasonally) males. Workers are wingless and do not reproduce, but they are often highly specialized for collecting food or defending the nest. Because workers are females, they have a stinging apparatus that is modified from the same type of ovipositor

Most ants have stingers, but this is not always the case. Wood ants, *Formica,* spray a solution of formic acid onto enemies and prey and then attack with their jaws.

used by the queen (a reproductive female) to lay her eggs. When the worker stings an intruder or an insect needed for food, she injects formic acid plus other chemicals produced by special glands above the sting. In most ants the venom is mild and the stinger is small, so they have no effect on humans other than producing an uncomfortable feeling. In some ants, however, there is a dangerous combination of potent venom, large stingers, and numerous aggressive workers that inflict many stings on a victim. These are the dangerous species.

Many tropical ants have gained a reputation for ferociousness that seems to be quite true. For instance, the bulldog ants of Australia (*Myrmecia*) are inch-long reddish species with large jaws in the workers. These workers defend the colony aggressively from any disturbance, rushing out to attack if even just a shadow falls across the nest. Dozens or even hundreds of stings can be inflicted on the unlucky victim in a matter of seconds, sometimes resulting in serious injury and even death. In Brazil, the almost equally large species of *Paraponera* are greatly feared. There may be up to 500 bluish black workers in a nest of these ants, and they are all vicious stingers. Even a few stings can cause sharp pain, swelling, chills, high fever, and an irregular heartbeat. The tropical army ants, *Eciton*, are feared not so much for their sting as for the tremendous numbers of workers that move through an area collecting all available food and protecting the column from all possible predators, including man.

One of the most feared of ants is the Australian bulldog ant, *Myrmecia gulosa* (below and below right). They have all the prerequisites that make an animal dangerous to man: they occur in large numbers in a colony, they react quickly and viciously to any threat, they are relatively large (1 inch), and their venom is very powerful and released in relatively large quantities.

The fire ants (*Solenopsis*) are greatly feared in the southeastern United States, and with good reason. Their bites cause great pain, lingering pustules and rashes, and often have serious side effects. When they attack an animal such as a calf in large numbers, they may actually kill it. Shown above is a slide preparation of the common red fire ant; shown below is its typical mound. Photo above by the Illustration Dept., Indiana Univ. Medical Center; that below by Dr. Sherman A. Minton.

In the United States only two types of ants are usually considered as a health hazard, in both cases because they form large colonies with aggressive workers that have potent stings. The introduced fire ants, species of *Solenopsis* (especially the so-called Argentine or red fire ant), build irregular mounds of dirt in open fields and near human habitations. These nests may be 2 or 3 feet high and may contain many thousand workers, each capable of repeatedly producing a very painful sting. Each sting results in a small blister surrounded by a reddish area, the blister breaking in a few days to leave a small pinhead-sized scar. Many people show various allergic reactions to the stings, and when in large enough quantity they may cause symptoms ranging from nausea and rashes to fever and respiratory difficulty. Deaths are rare, but they do occur if the allergic reaction is severe enough and there have been a great many stings.

The other seriously venomous U.S. ants are the various harvester ants of the genus *Pogonomyrmex*, aggressive species found mostly in the western U.S. and Mexico south into South America but with one species in the southeastern states. These relatively large ants have large spines on the body and build nests that are almost flat except for a low "collar" around the entrance and usually a wide cleared area around the nest. Although some species are gentle and seldom sting without provocation, others pour from the nest on little provocation and inflict dozens of painful stings in seconds. Other than intense pain and minor swelling, the stings have little lasting effect. Allergic reactions are possible, however, much as in fire ant stings.

WASPS

Wasps are important venomous animals throughout the world, especially as many types are social insects that occur in colonies of dozens or even hundreds of sterile workers (females) along with a queen or two and sometimes males. Many species build elaborate nests in the ground, in trees, or under the eaves or in the walls of houses, and many species are commonly found near human habitations. Wasps are readily recognized by their usually smooth and often glossy bodies, with the large abdomen usually separated from the thorax by a slender petiole. Actually, the distinction between bees and wasps is difficult to make since there are several rather different types of animals under each general category—usually bees are hairy and wasps smooth, but there are exceptions in each group.

The first type of wasp of medical interest is an exception to all the normal wasp traits we have mentioned. The body is covered with reddish and black hair, the wings are absent

The black and yellow wasps of the genus *Vespula* and their allies are familiar sights to everyone in the Northern Hemisphere. The many very similar species often build their nests in or near human habitations and are often attracted to spoiling fruit and garbage. Although their stings cause minor pain and swelling in most people, many deaths from allergic reactions have been recorded.

in the females, and they are not social. These are the so-called velvet ants, various genera of the family Mutillidae. Commonly called cow-killers and other interesting names, they are not ants but are actually wasps that parasitize beetle larvae and other insects. The females are often seen running rapidly along beaches and dry roadsides searching for buried beetle larvae in which to lay their eggs. The major danger from velvet ants is that they can be stepped on accidentally with bare feet or wander into clothing or shoes left on the ground. Their sting is immensely painful, sometimes almost incapacitating, but there are seldom any aftereffects.

Most other wasps that are of importance in human envenomation belong to either the family Vespidae (the great majority) or the Sphecidae. The sphecids are usually solitary wasps that nest in the ground, in hollow stems, or in mud nests. Most prey on insects and spiders, often paralyzing the prey and then laying an egg on it, the larvae

All wasps should be considered at least potentially dangerous, although many will sting only when severely provoked. Large wasps such as the tarantula killer shown here can cause very painful stings but seldom have any side effects.

hatching out and then feeding on the still-living food insect. The most familiar sphecids are various mud-daubers that build nests of mud tubes and provision spiders; spider wasps that attack large spiders in their burrows; and the cicada-killers that attack a variety of large insects. Most sphecids are reluctant to sting, but when they do, the sting is very painful.

The most important wasps are the social species of Vespidae. There are many types, including polistid paper wasps, hornets and yellowjackets, and tropical polybias. The many species of *Polistes* are usually slender, brown and yellow or brown and red wasps with long abdomens and a long petiole. They build circular paper nests with many cells in a single plane, the nest hanging from a nearly central stem. The cells face down and are open until the larvae develop into pupae, when they are capped with a convex white seal. Most *Polistes* nests are under 6 inches in diameter, but much larger nests are occasionally found. In the southern United States these wasps are active during much of the year and are especially dangerous because they tend to build many nests under eaves of houses. They are very painful stingers, but other than some swelling there is little other effect.

The tropical polybias, genera *Polybia*, *Gymnopolybia*, and others, are much like polistids both in appearance and habits. Their nests are usually rather conical and hang from tree limbs. The nests of polybias may be over 3 feet in diameter and contain thousands of workers. Disturbing a nest of these wasps is a painful experience that can occasionally be deadly if enough stings are suffered.

Home remedies seldom have any real effect on wasp and bee stings as they are based on the misconception that stings are injections of acids (actually they are alkalis). Prepackaged swabs containing chemicals that soothe the pain and reduce the swelling from common insect stings are available at most drugstores.

In the cooler climates of North America and Europe the most familiar wasps are the hornets and yellowjackets, species of such genera as *Vespa* and *Vespula*. These are usually black and yellow wasps with broad abdomens connected to the thorax by a short and not very distinct petiole. Identification of the various species is difficult, but recognition of the group is simple. It is best to stay away from any hornet or yellowjacket seen, as their nests are frequently hidden in the ground or between walls of buildings and might be disturbed in an effort to kill a single flying worker. When the nest is disturbed hundreds of workers may fly out and aggressively chase the disturber for many yards. Their stings are very painful and in quantity may cause severe allergic reactions in adult humans. These wasps are responsible for several deaths each year in the United States. Unlike the nests of polistids, the nests of hornets and yellowjackets are many-layered and often covered with another coating of rough paper, leaving only a single opening. These are the football-shaped nests usually thought of as typical wasp nests. The paper used in nest-making by these and other wasps is produced by repeatedly chewing small plant fragments.

Hornets (*Vespa*) are large wasps that often have the face solid yellow or white in color. Their football-shaped paper nests are familiar sights in many areas and are astutely avoided by the intelligent outdoorsman.

BEES

Without doubt, honeybees are the most deadly animals to be encountered in the United States and other cool northern countries. The single species, *Apis mellifera*, is responsible for about 25% of all deaths from animal envenomations in the United States, almost as many deaths as result from bites of all venomous snakes combined.

Everyone knows the honeybee. The orange and black workers can be found anywhere there are flowers, from Canada to the Southwest in the U.S. and in virtually all countries. Originally natives of Europe and Asia, honeybees were domesticated many centuries ago and spread with man. Beekeepers can be found everywhere, from the deserts to the largest cities, and their bees are perhaps the most familiar single species of insect. Although bees are not vicious stingers under normal circumstances, accidents may happen frequently almost anywhere, with people accidentally handling or stepping on bees or having their intentions misinterpreted by colonies. The result is one or more stings, and occasionally dozens or even hundreds of stings.

Honeybees are social insects, perhaps the most highly developed of any commonly seen social insects. Each hive has many workers, one large queen that lays the fertile eggs, and several short-lived males that are produced seasonally to mate with newly emerged queens. The workers, sterile females, determine the gender and status of the bees in the colony by controlling the food taken by the larvae. When new queens emerge from their pupal chambers, they commonly leave the hive accompanied by some of the workers and found their own nest. Honeybees communicate with each other through visual clues involving elaborate dances and also chemically through release of odors that cause the other workers to come to the aid of a threatened worker.

Usually the sting of a honeybee is a minor event resulting in mild pain and swelling that go away in a few hours or even minutes. However, many adults develop allergic reactions to beestings that vary greatly with the individual. At best such a reaction may include a rash, difficulty in breathing, and a low fever, all lasting just a few hours or a day. At worst, it results in death in minutes to days after a sting. Usually the most severe reactions occur after the victim suffers numerous stings, but deaths from a single sting are not uncommon. Also, adults are more likely to suffer severe reactions than are children, resulting in the idea that the reaction is due to sensitization—after having 30 or more years to accumulate beestings, the body becomes sensitive enough that just one more sting triggers the reaction. However, there are many reports of deaths in people who had not been stung often in the past; additionally, beekeepers claim that by being repeatedly stung they develop immunity to stings, not greater sensitivity. Regardless, severe allergic reactions in adult humans are not uncommon and may occur without warning.

Since worker honeybees have a barbed stinger, they can sting only one time. The sting is then left in the skin of the victim, and the worker dies because by trying to pull out the stinger she also pulls out much of her other abdominal organs. The glands at the base of the sting can continue to release venom for several minutes, so the stinger must be removed as soon as possible, preferably without pinching the glands. The pain is usually controllable by many different home remedies and chemicals to be found at any drugstore or supermarket, but allergic reactions require immediate medical attention.

There has been much publicity given to the so-called "African killer bees" that escaped from a Brazilian apiculture experiment in 1956. This bee, *A. m. adansonii*, an African subspecies of the honeybee, is almost indistinguishable morphologically from other varieties of honeybees cultured in the Americas, but it behaves quite differently. The workers are much more aggressive and

protective of the hive than most others, and they will attack continually. Apparently a worker will give off an odor when she is attacked or killed, and this incites other nearby workers to attack. As long as workers are attacking and being killed, they give off the chemical that spurs other workers to continue attacking. Victims that accidentally disturb a hive or kill a worker without thinking may receive hundreds of stings and die within minutes or days of the attack. Farm animals are also killed in this way, even cattle and horses. Between 1956 and 1976, over 150 people are known to have died in Brazil from beestings caused by this variety. (Actually, since this averages out to less than eight deaths per year in a country as large as Brazil, this number is not as tremendous as it first appears.)

Additionally, the African bees tend to form swarms of new queens more readily than other varieties and the swarms tend to take over other beehives, killing the less aggressive bees of other varieties. Swarms also commonly anchor themselves near human habitation, which is a major factor leading to human deaths. Fortunately, the African bees readily intermate with other varieties, which to some extent dilutes their aggressive behavior. If proper precautions are taken, they prove very valuable in controlled beekeeping in the tropics, where other honeybee varieties do not do well.

There is no doubt that African bees are moving north from their original point of release in Sao Paulo, Brazil. By 1975 they had spread over much of eastern South America, from Surinam to northern Argentina. In 1982 they were recorded in Panama, their first entry into Central America. At their current rate of spread they should reach the extreme southern United States by 1988. Their major effect in the U.S. will be not their venom but rather the impact they have on beekeeping practices as they mate with other strains of honeybees and force changes in established beekeeping procedures.

In addition to honeybees, bumblebees (*Bombus*) occasionally sting humans, but these large black and yellow hairy bees are seldom aggressive. Their colonies in the ground may contain several dozen workers, but usually they try to stay out of the way of humans. Bumblebees sting repeatedly and their sting is very painful, but allergic reactions are few. The same can be said for stings from the other types of bees encountered around the world—most species are small, nonaggressive, and occur in relatively small colonies. However, if any unusual symptoms develop after the sting of any type of bee or wasp, medical attention should be sought at once.

Facing page: The fuzzy black and yellow bodies of bumblebees are recognized by almost everyone, and almost everyone shows some fear of being stung by a bumblebee. Although the numerous species are often common and may build their burrows in flower gardens or along sidewalks, they seldom sting unless provoked. Fortunately, although their stings may be very painful, they seem to seldom cause allergic reactions.

10: Amphibians

Although we earlier promised to restrict this book to animals that not only produce a venom but have some way of injecting it into the victim's body, we will deviate a bit from this to say a few words about the frogs, toads, and salamanders, the familiar amphibians. Almost all of these animals produce toxic skin secretions and even more toxic chemicals in special parotid glands near the head or on the legs, but none have the ability to inoculate the poison; instead, they are harmful if mouthed or roughly handled.

Amphibians are, of course, familiar to everyone and are found throughout the world. The living members of the class fall into three groups: the common frogs and toads; the less familiar but often common salamanders and newts (absent from much of Africa and all of Australia); and the poorly known caecilians or apodans, legless worm-like amphibians of tropical America, Africa, and Asia. Although almost any amphibian can cause discomfort if its skin secretions are rubbed into an eye or placed on the lips, several species of true toads (Bufonidae) and their allies can cause severe poisoning or even death in humans.

Toads are frogs with special modifications for surviving in relatively dry habitats. The skin is often covered with warts, indicating the position of secretory glands, and large parotid glands may be present behind the head or on the legs. Various types of chemicals are secreted by these glands, but all have effects on the nervous system and can cause variations in heartbeat and blood pressure. In the larger species of toads such as *Bufo marinus* (introduced worldwide in tropical regions), *Bufo paracnemis* (a 9-inch species from southern Brazil, Paraguay, and northern Argentina), *Bufo arenarum* (Argentina), and *Bufo alvarius* (southwestern United States) the secretions of the glands may be both voluminous and strongly toxic, causing respiratory paralysis and eventual death in animals as large as dogs. When mouthed by a dog, the toad secretes its

The Colorado River toad, *Bufo alvarius,* is a 6-inch species found in the southwestern United States and adjacent Mexico. The parotid glands are especially prominent in this species and can be easily seen behind the eye and on the joints of the hind leg. Photo by James K. Langhammer.

Facing page: Even common garden toads, in this case the European *Bufo bufo,* can cause discomfort if mishandled and their secretions accidentally touch mucous membrane tissues.

Although it cannot inject its venom, *Bufo marinus* has long been considered at least potentially capable of causing human deaths. Certainly any small child that attempted to mouth one of these toads would be just as subject to the effects of its venom as a dog of equal weight—and dogs have certainly been killed. In some areas of Florida this is now a common species of toad.

poison, which rapidly enters the dog's bloodstream through the absorptive tissues of the roof of the mouth. In seconds the dog releases the toad and starts to clean its mouth, and in minutes the limbs weaken and fail. Death can occur in an hour or two. Of course humans are larger than dogs and do not usually put live toads in their mouths, but human children are closer to dogs in both size and curiosity, and accidents have happened, occasionally with fatal results.

The most toxic amphibians are perhaps the most colorful as well. These are the dendrobatid or poison arrow frogs of Central and South America. Close relatives of true toads, they are small (usually about an inch or so long), smooth-skinned, and often brilliantly colored in red, black, yellow, green, and even blue. Several species of the genus *Dendrobates* are utilized by natives to poison arrows. These totally unaggressive little frogs usually hide during the daytime in foliage, but various tribes of Indians have learned to collect them in quantity. The frogs are then placed over a fire or hot coals, either in a container or impaled on a sharp stick, and roasted until they produce a rather milky secretion from the skin glands. This secretion

The beauty of the poison arrow frogs makes them of interest to terrarium keepers, and several species appear commonly on the pet market. Shown here are *Dendrobates lehmanni* (top right), *D. auratus* (bottom left), and *D. histrionicus* (bottom right). Photo of *D. lehmanni* by W. Mudrack; that of *D. auratus* by George Dibley; and that of *D. histrionicus* by Aaron Norman.

Care should be exercised in handling any poison arrow frog. Several species have bred in terraria and may eventually become common pets. Fortunately the species are seldom dangerous even if badly mishandled. Shown are *Dendrobates pumilo* (top left), *Phyllobates bicolor* (top right), and *Dendrobates quinquevittatus* (bottom left). Photo of *D. pumilo* by George Dibley; that of *P. bicolor* by Aaron Norman; and that of *D. quinquevittatus* courtesy of Dr. D. Terver, Nancy Aquarium.

Several newts have been shown by laboratory tests to have very potent skin secretions, but since they cannot inject the venom they are of little or no importance in human medicine. Photo at right of the western American *Taricha torosa* by Aaron Norman; that below of the European *Triturus cristatus.*

contains the toxin. It is scraped off by the Indians and rubbed on arrows to produce extremely deadly weapons; often the frog venom is mixed with herbs or decaying animal matter to increase its effectiveness. Under normal circumstances, even if roughly handled dendrobatids seldom secrete their toxin, so the origin of the Indians' knowledge about how to secure it remains something of a mystery.

Among the salamanders much the same situation occurs as in the frogs—all species probably produce skin secretions, but unless rubbed in the eyes or on vascularized tissue the toxin is harmless. The larger salamanders such as the mudpuppies, *Necturus,* can produce a goodly amount of toxin of some potency, but as they have no method of injection it will only discourage predators that mouth the animal. Newts, Salamandridae, have several especially toxic species, with the western American *Taricha* producing a toxin capable of killing mice almost instantly if injected. However, unless you intend to eat salamanders alive and with their skin, they present no danger to humans.

11: Gila Monsters and Beaded Lizards

Although many people consider lizards such as skinks and anoles to be venomous, with only two exceptions lizards are totally devoid of both venom glands and the teeth with which to inoculate the poison. Although accidental bites from larger lizards may be painful and may even draw blood, they are harmless (though secondary infections may develop in an uncleaned wound).

The two species of the family Helodermatidae are the exceptions to this, and their undoubted venomosity combined with their unusual and rather "primitive" appearance have caused them to enter the realm of animal legends.

The gila monster (pronounced "heela"), *Heloderma suspectum,* derives its name from the Gila River system of southwestern Arizona. It is fairly common in the drier regions of Arizona, New Mexico, and northern and northwestern Mexico, and possibly occurs also in areas of Utah and California bordering the main range. It is a stout lizard with a large head, strongly clawed legs, and a heavy tail; it may reach a total length of about 18 to 24 inches. The dark brown to black body is variably mottled or banded with pink and white, and the bead-like scales have bony cores. The tail varies considerably in size because it is used to store fat against adverse times when food and water are scarce.

The Mexican beaded lizard, *Heloderma horridum,* is larger (24 to 30 inches) and usually darker and less brightly colored than its more northern cousin. Its range is restricted to Mexico, where it is found from the northeastern shore of the Gulf of California south along the Pacific coastal deserts to the northern edge of the tropical rain forests (which it does not enter). The tail is longer and usually more slender than in the gila monster and may be a bit prehensile; there is evidence that the animal commonly seeks cover by climbing in low shrubs.

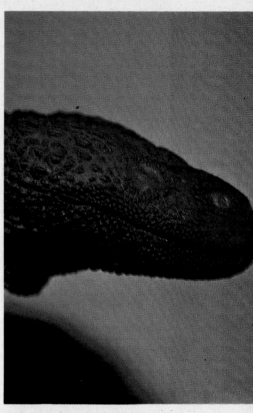

The rare Bornean earless monitor, *Lanthanotus borneensis,* was once thought to be closely related to the gila monster and thus at least potentially venomous. More detailed study of fresh and living specimens shows this idea to be erroneous, however, and there are still just two known species of venomous lizards living today. Photo by J. T. Collins.

Facing page, above and below: Most populations of gila monster, *Heloderma suspectum,* are brightly mottled or blotched with pink, orange, or white. Because of their large size, docile temperament, attractive appearance, and—of course—their reputation as deadly animals, gila monsters are very popular exhibit animals in both zoos and small roadside displays. This has caused a great strain on the wild populations. Photo at top by H. Hansen, Aquarium Berlin.

Both species feed largely at night and seem to prefer raiding nests of birds and small mammals, eating the eggs and young. In captivity they do quite well on a mixture of raw chicken eggs and chopped beef. When attacking a whole egg, the lizard pushes it around until it finds a point of support, then it breaks the shell and licks out the contents with

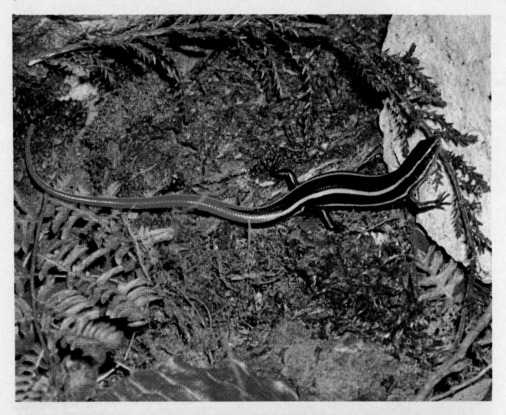

For some reason, perhaps their shiny appearance or their often brightly colored heads or tails, skinks (family Scincidae) are widely believed to be venomous. There is of course no validity to this idea, but it is often hard to convince people of this. At the top is *Eumeces skiltonianus*, photo by Bertrand E. Baur; at the bottom is *Eumeces laticeps*, photo by F. J. Dodd, Jr.

The Mexican beaded lizard, *Heloderma horridum,* is very similar to the gila monster but has stronger legs and a somewhat more elongated body and tail. The habits of the two species are very similar, however, and both are considered to be about equally venomous.

its wide, fleshy tongue; the last dregs of yolk are obtained by turning the broken shell over. Prey is found largely through smell and use of the tongue, as eyesight is weak.

Four to 15 eggs are deposited in a shallow nest. Reproduction is poorly known, but an old record indicates the young hatch after an incubation period of about 30 days, which some authorities think is too short to be correct. The young are about 6 inches long at hatching. Specimens have lived at least 20 years in captivity.

Unlike other venomous reptiles, the venom glands of these lizards are modified salivary glands located in each *lower* jaw. Each gland is formed from several different sections, each with a separate duct leading to the bases of grooved teeth in the lower jaw. Thus there is no single fang as such, but instead a small battery of teeth modified for injecting poison into the wound produced by steady chewing. Even some of the upper teeth are grooved, although they do not have venom ducts at their bases. When biting, the gila monster or beaded lizard bites quickly for an animal of its size and apparent sluggishness and hangs on, chewing continually. A fair amount of venomous saliva is thus inoculated into the wound. Although the jaws grip strongly, the teeth are easily broken at their bases (they are not in sockets) and it is relatively easy to *pull* the animal off rather than trying to break its grip. The venom causes a sharp pain, swelling, heavy sweating, and sometimes fainting. It is a weak neurotoxin and can thus cause respiratory failure, but human deaths are very rare (about eight recorded in 100 years). Although gila monsters can react very quickly (they are not nearly as sluggish as they appear), they will take quite a bit of abuse before biting. Almost all bites result from prodding captive or confined animals with a finger or bare foot.

Although many herpetologists of even just a century ago adamantly refused to believe that gila monsters could be dangerous to humans, the Spanish explorer and adventurer Francisco Hernandez wrote a rather accurate report on its toxicity as early as 1577: "Though harmful, the bite of this animal is not deadly; the beast is feared more on account of its appearance than for the effect of its bite, and in fact it does not attack man unless offended or aroused." This sentence could accurately summarize the results of more recent studies. Today man is more harmful to gila monsters and Mexican beaded lizards than vice versa, as both species are popular exhibition animals and are heavily collected. Arizona has made it illegal to collect gila monsters, and eventually both species will certainly be considered as threatened by human activity.

12: Introduction to Venomous Snakes

The very mention of the word "snake" sends some people into hysterics, and many people cannot stand to look at even a photo of a snake in a book. Although many writers and perhaps the great majority of everyday people believe that this dislike of snakes is a natural human trait, this is simply not so. The fear or hatred of snakes is taught to children by their parents and peers, often with religious overtones—afterall, most Western religions feature the snake as evil in one way or another, and everyone "knows" that snakes are deadly poisonous. Unindoctrinated children are not afraid of snakes and will attempt to handle any snake just like an interesting plaything (which has unfortunately resulted in death on several occasions when the wrong snake was handled). Many adults also enjoy handling snakes of various species. Certainly there is no natural fear of snakes.

However, unless children and adults are instructed as to which snakes are harmless (the great majority of species) and which are venomous (only a few in any country), perhaps it is best that most people remain afraid of snakes to some extent. If a person is not willing or able to learn to distinguish poisonous species, then it is best to treat all snakes as venomous and leave them strictly alone and unmolested. Most incidents of snakebite, at least in more industrialized countries, occur when a snake is handled or attacked, with smaller numbers occurring from accidentally stepping on snakes or threatening them unintentionally.

Although the next several chapters will focus on venomous snakes and their effects on man, naturally emphasizing the danger of snakebite, it should be emphatically stated that most snakebites do not cause death in adults even if untreated. Most venomous snakes have only a 10 to 20% death rate in untreated bites, although a few species have

Although copperheads are abundant in many parts of the eastern United States, they seldom cause fatal or even serious accidents. *Agkistrodon contortrix* fortunately is a species that has a relatively weak venom, for it has very effective camouflage colors and is thus more likely to be stepped on. Photo by Jeff Gee.

Facing page: At the other end of the scale are the mambas, *Dendroaspis*, very large snakes with long fangs and extremely potent venom. Bites from mambas often happen far from hospital facilities, often are situated near the head or chest (because of the tree-dwelling habits of the snakes), and are often rapidly fatal.

Bitis peringueyi from southwestern Africa is the epitome of the general conception of a venomous snake—broad head, narrow neck, stout body, short tail, dull colors. Unfortunately, few venomous snakes have all these features, and many have none. Photo by P. van den Elzen, courtesy Dr. D. Terver, Nancy Aquarium.

higher rates and many species have much lower rates. The copperhead of eastern North America, for instance, is perhaps the most common venomous snake in its part of the world and causes many bites each year, yet its bite is seldom fatal even if untreated, assuming the victim is an adult in good health and the bite does not occur in an unusual location such as the chest. The venom apparatus of snakes is primarily a device for obtaining food, and the venom tends to be most effective against the normal prey of the snake species. Thus the bite of a copperhead (or cobra for that matter) is adapted for bringing down small mammalian and bird prey. It works best on rats or sparrows and is injected in quantities suitable for killing an animal the size of a rat. It is also not uncommon to find snakes that have recently fed and have to a large extent depleted the venom in the poison sacs. Some species of venomous snakes (such as the sea snakes) have a reputation for biting without delivering a dose of venom. Many venomous snakes are of relatively small size and have short fangs, with the result that their bite is easily deflected by normal clothing and shoes.

In India, a heavily populated, highly agricultural, and medically relatively primitive country, the 50-plus species of venomous snakes probably cause about 6,000 to 9,000 deaths a year, of which 90% represent victims who received either no treatment or "home remedies" that did no good. Of the about 25,000 Indian snakebite victims who enter government hospitals each year, only 5% die, and most of these entered the hospital one or two days after the bite, when it was too late for effective treatment. The overall snakebite mortality in India is probably in the vicinity of 10 to 20%, mostly due to lack of adequate treatment. If more antivenins and supportive treatments were available, the total mortality rate would probably drop to about 5 to 10%. And this is in a country with cobras, kraits, Russell's vipers, and saw-scaled vipers, where cobras are tolerated in houses, and where agricultural workers go about without shoes and with the legs often bare! In more advanced countries death from snakebite is so rare that almost every case becomes a newspaper story and may receive mention on radio and television. So never assume that if you are bitten you are dead, because it is just not so.

Many nonvenomous snakes, however, do have many of the features usually thought of as those of venomous species. Just compare this photo of the common (harmless) hognose snake, *Heterodon platyrhinos,* with the picture of the *Bitis* on the facing page. Photo by James K. Langhammer.

VENOMS

Snake venoms are complex mixtures of proteins produced and stored in poison glands at the back of the head. The glands are similar to salivary glands and may be small and not very obvious, as in coral snakes, or large and readily obvious, as when they form the triangular head shape typical of pit vipers. In all venomous snakes the glands open through ducts into grooved or hollow teeth in the upper jaw. Even in some nonvenomous snakes, however, the saliva is suspected of having an effect on prey animals almost equivalent to being an actual venom. Thus there are several reports of rather typical (though mild) venom reactions in humans after the bites of ordinarily nonvenomous snakes such as species of water snakes and hognose snakes, both of which lack grooved teeth and venom glands in the normal sense of the word.

The proteins that make up the venom are of two types, each distinguishable by its activity, with several subtypes. One category of venoms is the neurotoxins. These venoms affect the nervous system, causing destruction or paralysis of the nerves that regulate heartbeat and respiration. Eventually the victim may die from asphyxiation or heart failure.

The other major group is the hemotoxic venoms. These attack the blood cells in several ways and also destroy tissue, both muscular and vascular. The lining of the blood vessels is destroyed by hemorrhagins, allowing blood to escape into the body cavity; red blood cells are destroyed by hemolysins; thrombases cause clotting of the blood within the vessels, while anticoagulants prevent blood from clotting; cytolysins destroy any cells they come in contact with, whether blood or muscle.

In practical terms, however, the situation is far from being so simple. Since venoms are a mixture of proteins, they commonly contain both neurotoxins and hemotoxins. The proportion of these types varies from genus to genus and even from species to species, making generalizations difficult and often misleading. Thus, it is fairly safe to say that cobras are neurotoxic, but the closely related mambas are partially hemotoxic and the several mamba species each have different venoms. Among the pit vipers most are decidedly hemotoxic, yet even within certain species of rattlesnakes one subspecies can have high amounts of neurotoxins in its venom and produce very different symptoms and effects than another almost identical subspecies. As a general rule, however, most Elapidae are neurotoxic (mambas being notable exceptions) while Viperidae are hemotoxic (with several exceptions among several genera).

FANGS AND CLASSIFICATION

The venom glands are connected by ducts to the bases of the fangs, elongated teeth in the upper jaw. Among venomous snakes these teeth vary a great deal in size and in details of shape, as well as in placement. Before the current classifications of venomous snakes became more complicated (and probably more accurate), snakes were divided into four groups based on the structure of their teeth: Aglypha, harmless snakes with solid teeth; Opisthoglypha, with grooved teeth, usually at the back of the mouth (rear-fangs); Proteroglypha, with hollow immovable fangs showing traces of grooves at the front of the mouth; and Solenoglypha, with hollow movable fangs without grooves. These groups were supposed to represent a gradual evolution from harmless snakes to mildly venomous snakes with poorly developed and badly placed teeth, and then to advanced and fully developed highly venomous snakes with perfected teeth.

Unfortunately, as more knowledge accumulated about snake morphology, it readily became apparent that this simple classification would not work. Today there is more doubt than ever as to the relations of the venomous snakes, with a growing tendency to believe that the cobra and viper groups are composites consisting of several types of unrelated snakes that have developed similar venom apparatuses along parallel lines. The old concept of rear-fangs as a group of related snakes was demolished long ago when it was shown that unrelated snakes in the Old World and

One of the more dangerous small (3 feet or so) rattlesnakes of the western United States is *Crotalus viridis*, a species that occurs in a confusing array of patterns. Although its bite is seldom fatal, it often causes severe tissue damage and scarring.

124

New World had both evolved into similar rear-fang types with almost identical teeth and head structures.

For our purposes, venomous snakes can be divided into three families, each with subfamilies. The rear-fanged snakes such as the boomslang, vine snakes, and cat-eyed snakes belong to various subfamilies (formerly all in a subfamily Boiginae) of the large snake family Colubridae, comprised with few exceptions of nonvenomous snakes and mildly venomous snakes harmless to man. The cobras, coral snakes, sea snakes, and most Australian snakes belong to the family Elapidae. The sea snakes are extremely modified cobras with adaptations for living in the sea; they are placed in the subfamily (formerly considered a full family) Hydrophidinae. The New World coral snakes are now considered by some authorities to be unrelated to the Old World cobras and to have developed independently in the Americas; for this reason they are sometimes placed in the subfamily Micrurinae (considered by some authorities to be a full family). The vipers (Viperidae) are even more complicated. There are two basic subgroups recognized, the true vipers without a heat-sensing pit between the eye and the nostril (Viperinae) and the pit vipers with a heat-sensing pit (Crotalinae). The Viperinae are exclusively Old World snakes found through much of Europe, all of Africa, and almost all of Asia. Certain odd vipers from Africa recently have been moved into separate subfamilies because they are no longer thought to be closely related to the other vipers. The pit vipers (Crotalinae) are somewhat unusual in that they occur in both the New World and Asia, with one genus having species in both these areas.

In the rear-fangs the modified teeth are at the posterior end or near the posterior end of the maxillary bones. (In snakes there are four rows of teeth on the upper jaws, with the outermost row on each side a maxillary row.) Their fangs are usually shallowly to deeply grooved and are commonly separated from the more anterior teeth (if there are any) by a gap or diastema. Muscles can cause the back end of the maxillary to move down and forward somewhat when the snake strikes, but usually the snake must chew in order to get the hind teeth into play and inject venom. Since many rear-fangs are small or have small gapes, accidental envenomations seldom occur in man unless a snake is being handled and is allowed to bite and continue chewing for awhile.

In the cobras and their allies (elapids) the fangs are at or near the front of the mouth and the groove has become deep enough that the teeth are actually hollow. In most elapids the edges of the groove close over so the duct inside

Several instances of harmless snakes causing moderately severe (allergic ?) reactions in humans following bites have been recorded. Even the water snakes and their relatives, such as this Eurasian *Natrix maura*, have caused such reactions, though their saliva would under normal circumstances not be considered dangerous. Photo courtesy Dr. D. Terver, Nancy Aquarium.

The African genus *Bitis* is greatly feared, and with good reason. The very length of the fangs allows these snakes to make serious bites in almost any animal and to penetrate even stout shoes and clothing.

Facing page: Sea snakes are in some ways the most unusual of snakes, although they are obviously close relatives of the cobras. *Laticauda* is one of the more primitive sea snakes and retains many similarities to its cobra relatives, such as large ventral scutes, a relatively normal head, and the ability to lay eggs on land.

the tooth is fully enclosed or nearly so (though still showing external traces of the groove), but in some it is more open. In spitting cobras the opening at the end of the fang is slightly above the tip so the jet of venom is directed outward from the mouth rather than downward. In elapids the fangs are relatively short even in large species, seldom exceeding a quarter of an inch. Usually they are straighter and not as curved as in the vipers, but this varies considerably. In the mambas the fangs are large, curved, and placed at the very front of the mouth much as in vipers. In the cobras the maxillary is almost immovable, so the fangs are always in the biting position; this is somewhat related to their shorter length, as in the mamba with its larger teeth the maxillary is almost as movable as in the vipers. The fang may be followed by other maxillary teeth in some genera of elapids or may be the only tooth on the bone.

In the vipers the fangs tend to be much longer than in most elapids of similar size, are curved, and are alone on the end of a short and very movable maxillary bone. The lumen of the tooth is entirely internal, with little or no trace of the original groove on the surface of the tooth. When not in use the fangs are usually folded back against the roof of the mouth as they are too long to allow the mouth to be closed with them erect. In the South African *Atractaspis* the fangs are so long in relation to the mouth that they are usually kept outside the mouth and the snake bites with the mouth closed. (This is not as simple as it sounds; the lower jaw is flexible and must do contortions to get out of the way of the fang.) In most smaller vipers the fangs are only a quarter of an inch or so long on the curve, but in some of the truly large vipers such as the Gaboon viper, bushmaster, and large diamondback rattlesnakes the fangs may exceed 1 inch in length and may approach 2 inches. With the large fangs go large venom glands with massive doses of venom being injected in each bite. These are truly the most dangerous of the snakes, even if laboratory tests show that the venom of many elapids is more potent when injected into mice. Elapids are simply not as efficient in the development of the venom apparatus as are the vipers.

FIRST AID AND AWARENESS

The question of first aid procedures for snakebite is a difficult subject. Over the years there have been many different suggestions on how best to cope with a venomous snakebite. Of course today everyone discourages such "home remedies" as drinking liberal doses of alcohol or using gunpowder on the bite, but beyond this there is little agreement. At one time almost every book, popular or scientific, suggested the cut-and-suck method, which consisted of making X-shaped cuts over each fang mark and sucking out the venom by mouth or with rubber bulbs. Because of the rapidity of the spread of venom, the chances of secondary infections, and the lack of evidence that the method works, the cut-and-suck method is no longer recommended (although some authorities say it works well with North American pit vipers). Cryotherapy is now somewhat in vogue, with recommendations that the bitten limb be immediately immersed in ice cubes or cold water. Some authorities, however, feel that cryotherapy does little good and do not recommend its use.

Frankly, first aid for many snakebites is probably harmless but also does little good. Because of the problem in recommending first aid measures, all we will suggest are the very basic steps of cleansing the wound, identifying the snake to make sure it is really venomous, and getting immediate medical attention and, if necessary, antivenin. If the bite is on a limb a tourniquet can be applied above the site to slow the spread of venom (if hemotoxic in nature), but tourniquets have their own problems and are not recommended by some doctors. Of course the value of a tourniquet in slowing the spread of neurotoxic venoms is probably nil. If a tourniquet is used, be sure to release it for half a minute to a minute every 20 to 30 minutes and do not keep it on for over two to three hours.

If at all possible, secure the body of the snake involved in the incident. This will ensure accurate identification. Not only is it important that the doctor know the bite was actually caused by a venomous species, but also in many areas of the world it is necessary to know just what species was involved in order to use the proper antivenin. (Looking for fang marks often will not work.) Antivenins for pit vipers will not work on cobra bites and vice versa, although there are polyvalent antivenins available in some areas that combine factors for several different species. If you were bitten by a harmless snake you do not want to receive antivenin as there is a good chance of reaction to the horse serum from which it is made. In fact, before you are given the antivenin a doctor will make a skin or eye reaction test to determine if you are allergic or not. Additionally, after receiving antivenin there is often a tendency for increased sensitivity to future doses of venom and antivenin. So try to get at least the head of the snake involved.

Even after antivenin is used, there is often the necessity for supportive care of some type because of nerve, blood vessel, or muscle damage as a side effect of the bite. Some snakes produce great tissue damage near the site of the bite,

In many vipers the pattern is a simple one of a row or rows of dark blotches down the middle of the back and along the sides, often with dark stripes running back from the eyes. The pattern of this Central American *Bothrops godmani* is in general similar to that of literally dozens of other vipers and pit vipers. Photo by John T. Kellnhauser.

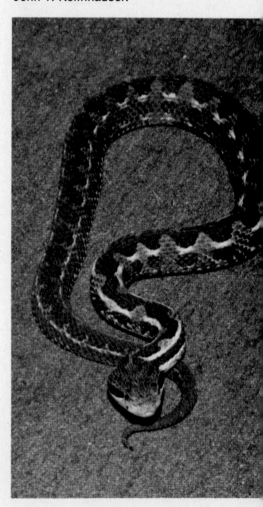

Patterns of bright red, yellow, and black bands have evolved in several groups of elapids, including the American coral snakes. In North and Central America several species of harmless kingsnakes (*Lampropeltis*) have developed similar banded patterns that are apparently accurate enough to at least confuse their enemies (including snake collectors). Photo of *Lampropeltis triangulum annulata*, the Mexican milk snake, by Ken Lucas, Steinhart Aquarium.

with scars that will remain visible for years. Without doubt the bite of any venomous snake, regardless of its size, should be given prompt medical attention.

Try to avoid scaring the victim. There are authenticated cases of people dying from the bites of harmless snakes due to heart attacks or shock, and some have even died because they *thought* they had been bitten—with no real snakes involved. Keep calm and remember that even if untreated very few snakebites are fatal. Most snakes are just not that poisonous, and many are not 100% effective at the time of the bite, having partially empty glands or making poor strikes with only one fang partially entering the victim.

In nonagricultural areas, the main causes of snakebite are probably attempts at handling venomous snakes either in public displays or private collections and misguided attempts to kill or torment wild venomous snakes. With only a few exceptions—and even these exceptions are seldom well documented—venomous snakes are not aggressive and will not strike unless they feel cornered. If you leave them alone, they will leave you alone.

Watch where you put your hands and feet when in an area known to be frequented by venomous snakes. Unless you step directly on a snake or put your hand on one, the odds are that the snake will retreat. When lifting objects, lift from behind, not from in front. Never poke your hand into a crevice or hollow log or under a ledge without checking it closely first (and *not* by sticking your head into the opening!).

Learn to recognize the venomous species in your area. Almost every country has at least one available book and usually several pamphlets on venomous local snakes. In most areas it is not difficult to recognize at least the genera of poisonous species. If you know which species are harmless and which venomous, it is easier to avoid accidents with the venomous ones.

Because of their low body weight, children are more apt to be severely injured in accidents with venomous snakes than are adults, and even mildly venomous species can be the source of serious bites in children. Children are also more likely to pick up a snake out of curiosity than are adults; in fact, several of the few fatalities from bites by eastern North American coral snakes have resulted from children picking up the brightly colored animals and treating them like toys. Either teach children how to recognize and respect venomous local species or teach them to avoid all snakes and just leave them all alone.

13: Rear-fangs

The snakes here termed rear-fangs are a miscellaneous assemblage of probably unrelated colubrid snakes. What they have in common is the possession of a pair of at least somewhat enlarged teeth at the rear of the maxillary bone and grooves on these teeth. Once they were all placed in a single subfamily, Boiginae, defined by these characters. Now, however, the Boiginae is more or less restricted to close allies of the mangrove or cat snakes of Asia, the genus *Boiga,* and the various rear-fang genera are placed in several subfamilies or generic groups.

Because rear-fangs tend to be small in size, have a relatively mild (to humans) venom, and have to chew for several seconds or even minutes to get venom into the wound, they are usually of minor importance in human snakebite. In the United States, for instance, there are several genera and species of rear-fanged snakes found mostly in the southwestern part of the country. All are uncommon and highly valued by collectors, with the result that they are commonly mishandled and even allowed to bite fingers just to see what their bite is like. So far there have been no serious injuries from such voluntary bites, with at most local swelling and some pain that soon goes away.

The tropical rear-fangs that enter the U.S. from Mexico include such genera as *Leptodeira, Coniophanes, Trimorphodon,* and *Oxybelis,* all typically Central and South American types. *Oxybelis,* the vine snakes, probably have the worst bite so far recorded, producing considerable swelling of the bitten finger and pain about like a beesting. These very thin, elongate, long-headed brown or green snakes are highly arboreal (tree-dwelling) and seldom are found on the ground. In Asia the genus *Ahaetulla* is almost identical in shape, as are the genera *Leptophis* (South America) and *Philodryas* (South America), which are also mildly venomous rear-fangs.

This tropical American vine snake, *Oxybelis fulgidus,* has a mildly potent venom and short posteriorly placed fangs. It must chew its prey for up to several minutes to inject sufficient venom to kill it. Photo by Dr. Marcos Freiberg.

Facing page: The plains of southern South America are inhabited by several species of mostly greenish, elongated snakes of the genus *Philodryas* that are rear-fanged and mildly venomous. Shown are, from top to bottom, *Philodryas baroni, P. aestivus,* and *P. olfersii.* Photos by Dr. Marcos Freiberg.

An array of commonly seen rear-fangs. **Above left:** *Leptodeira annulata* (America); photo by John T. Kellnhauser. **Above right:** *Boiga irregularis* (Asia); photo by John T. Kellnhauser. **Below:** *Hypsiglena torquata* (western North America); photo by Ken Lucas, Steinhart Aquarium.

Many of the rear-fangs would be harmless to man even if their venom were quite potent because their fangs are simply too small to break the skin even if the snakes were allowed to chew. Many of the rear-fangs are rarely seen in captivity, and many are also uncommon in nature. Shown here are, top to bottom: *Elapomorphus bilineatus* (southern South America), photo by Dr. Marcos Freiberg; *Tantilla nigriceps* (western United States), photo by Richard L. Holland; *Philodryas patagoniensis* (southern South America), photo by Lucrecia C. de Zolessi.

Within the rear-fanged snakes of even a small region such as southern South America, the head shape may vary tremendously. In *Philodryas baroni* variety *fusco-flavescens* (top) the head is elongated and the snout projects, much as in the vine snakes; photo by Dr. Marcos Freiberg. *Thamnodynastes strigatus* (center) has a relatively narrow head little distinct from the neck; photo by Rogelio Gutierrez. When irritated, the head of *Tomodon dorsatus* (bottom) may appear just as wide and arrowhead-shaped as in any truly venomous viper; photo by Dr. Marcos Freiberg.

The Asian flying snakes (*Chrysopelea*) are among the few reptiles that can glide from tree to tree. This is done by spreading the ribs and producing a broad, concave ventral surface that acts to give a considerable amount of lift. They can glide easily from branch to branch far above the ground. Their bite is usually harmless to man.

Most rear-fanged snakes are tropical in range and many are nocturnal, feeding on lizards, small mammals, and baby birds. Many of the nocturnal species have vertical pupils, an adaptation unusual in temperate area snakes. Among the many genera not already mentioned are *Crotaphopeltis* (Africa), *Imantodes* (tropical America), *Oxyrhopus* (South

The coral snake pattern of red, black, and yellow bands has been imitated by many South American snakes, including a few rear-fanged species. Shown here are *Oxyrhopus rhombifer* at the top and *Oxyrhopus trigeminus* at the bottom, both species from Brazil and Argentina. Photos by Dr. Marcos Freiberg.

135

Left: *Chrysopelea ornata,* the Asian flying snake, can glide from tree to ground by spreading the belly and making the surface concave to provide more lift. Photo by G. Marcuse.

Below: Another arboreal rear-fang is *Imantodes cenchoa* of tropical America. The very long and thin body with especially thin neck and tail allows it to live and hunt on even very small branches. The large eye with a vertical pupil is typical of nocturnal or dusk-feeding rear-fangs as well as most vipers. Photo by John T. Kellnhauser.

The boomslang, *Dispholidus typus,* is the only rear-fang that has gained a reputation for being seriously venomous to man. Although it must be allowed to chew for a few seconds to inject its venom, it has caused several deaths, mostly among amateur and professional herpetologists carelessly handling the snakes.

America), *Psammodynastes* (Asia), *Psammophis* (Africa), *Tantilla* (eastern U.S. to South America), and *Thelotornis* (Africa). Any large specimens of these and related genera should be treated with caution as it is possible that some species could produce serious envenomation if allowed to bite and chew. *Thelotornis*, the bird snake, has indeed been indicted in several cases of serious poisoning in man.

The only deadly venomous rear-fang so far known is the boomslang, *Dispholidus typus,* of southern and tropical Africa. This 5-foot species is a slender, exceedingly swift tree-dweller that varies in color from greenish to brownish or reddish. When tormented it swells the throat and front part of the body to several times their normal diameter, a feature of behavior found in several unrelated tree snakes. Although it is slow to anger and must chew to inject its venom, it has caused at least a score of human deaths, including that of the famous herpetologist Karl P. Schmidt. Until Schmidt died in 1957, most herpetologists considered the boomslang to be either harmless or only mildly venomous. The venom is hemotoxic.

That the fear of snakes is not ingrained in children is shown by the behavior of children with coral snakes. They pick up the beautifully banded animals and let them chew on a finger, which is about the only way a coral snake can inject venom into a human. Photo of *Micruroides euryxanthus,* the Arizona coral snake, by James K. Langhammer.

14: Cobras and Their Allies

Elapidae is a large family of about 50 genera of cobras and coral snakes plus another 15 or so genera of sea snakes. Although all members of the family have fangs and venom, many are too small to bite a human, are totally nonaggressive, are very rarely seen in the wild, or have reputations for releasing extremely small amounts of venom when biting. Thus, of the almost 100 species of Elapidae known from Australia (where most of the snakes belong to this family), probably only five or six species consistently have been the causes of serious human envenomations, with another 15 to 20 species (mostly sea snakes) theoretically capable of serious bites. The remaining species are too small or too secretive to be considered a serious problem.

COBRAS

Cobras are the most famous elapid snakes. This rather diverse group of large, active, terrestrial (rarely aquatic) snakes from Africa and Asia is famous for such species as the spectacled or hooded cobra, *Naja naja;* the African banded cobra, *Naja haje;* the king cobra, *Ophiophagus hannah;* the ringhals or spitting cobra, *Hemachatus haemachatus;* and the African desert cobra, *Walterinnesia aegyptia.* All share the ability to rear the front part of the body high off the ground and have specialized ribs allowing the neck to spread into a hood. They are about 1 to 5 feet long except for the king cobra, which can reach or exceed 12 feet in length. All lay eggs except for the ringhals, which gives live birth (ringhals differ from the typical cobras in numerous other ways and are probably not really related). The king cobra female actually builds a nest for her eggs and curls around it to protect the eggs; she is the only snake known to do this. Though some pythons curl around the eggs to incubate them and other elapids guard their eggs, they do not build nests.

Facing page: The ringhals or spitting cobra, *Hemachatus haemachatus,* can effectively spray its venom a distance of as much as 7 feet and still hit a human eye. Although its preferred defense reaction is to spit, it can of course also bite if cornered or mishandled.

The hooded or spectacled cobra, *Naja naja,* is the only Asian member of the genus, the others being found in Africa. Mongooses have long been considered the major enemies of cobras (though this is almost certainly not the case in nature), and fights are still staged to see if the swift movements of the mongoose will allow it to kill and eat the cobra before the reverse happens. Mongooses are not immune to cobra venom.

The Egyptian cobra, *Naja haje* (not to be confused with the scientific name of the hooded cobra, *N. naja*), occurs in the drier areas of northern and central Africa and also in the Arabian Peninsula. It is the only African cobra that has a row of small scales under the eye (visible in close inspection of living animals). The hood lacks the pale markings of the hooded cobra. Photo by Ken Lucas, Steinhart Aquarium.

Cobra venom is neurotoxic, causing local pain (sometimes absent), swelling, blurred vision and drooping of the eyelids (ptosis), speech problems, increased salivation and drooling, respiratory distress, coma, and eventually death in up to 20% of the cases if untreated with antivenin. The venoms of the various species or subspecies of cobras differ considerably in toxicity, and in the spitting cobras (both *Hemachatus* and some species of *Naja*) the venom is considerably diluted to allow it to be sprayed. A bite from any cobra is serious, however.

The ringhals and various species and subspecies of African *Naja* are the spitting cobras of safari movies. The opening of the fang is slightly above the tip, allowing the venom to be sprayed or ejected in the direction the snake is facing. The spitting cobras seem to have the ability to aim very accurately at reflective surfaces, including human eyes. A large dose of venom in the eyes can cause temporary or permanent blindness. Flooding the eyes with large quantities of milk or vinegar has been suggested as the best emergency antidote for venom in the eyes.

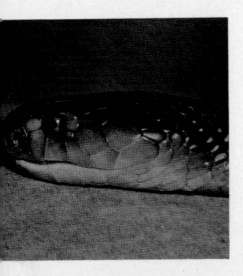

Two views of a hooded cobra from Taiwan, *Naja naja atra.* In this subspecies there is little development of the hood pattern, and the hood is barely distinguishable when the snake is relaxed. The large scales between the eye and the upper lip as shown in the detail view above are typical of most cobras. Photos courtesy Dr. R. E. Kuntz.

Two views of the many-banded krait, *Bungarus multicinctus.* Although at first glance this species may look like the common banded krait, it is readily distinguished by having a pointed tail (blunt in the banded krait). The species is often common in rice paddies. Photos courtesy Dr. R. E. Kuntz.

KRAITS

Kraits are members of the genus *Bungarus* found from the Indian area through Southeast Asia and into much of China. They are nocturnal, inhabiting burrows during the day and coming out at night to feed. Some species are typically found near water, others in drier savannahs. Snakes, small mammals, lizards, and even fishes are preferred foods of the various species. In India and Taiwan, kraits are often found near houses and even in cities, where they prey upon mice and rats.

Kraits have very toxic venom that can easily cause human deaths. Additionally, their bite is virtually painless and causes few local symptoms for the first three to five hours, when the typical symptoms of cobra poisoning develop. Treatment is thus often delayed until too late. Fortunately, kraits are also very gentle and nonaggressive animals that seldom bite even under extreme provocation. They are captured in large numbers by commercial snake dealers and handled with impunity, yet seldom with any accidents resulting. There is no doubt that they are very deadly snakes, however, if they do bite. They have strong jaws and tend to hang on tenaciously, chewing to inject more venom.

For a venomous species that may occasionally reach 7 feet in length, the banded krait, *Bungarus fasciatus,* is in actuality virtually harmless, and many natives in Southeast Asia and India will not believe it is venomous. It is very shy and almost never bites. Photo by G. Marcuse.

The Oriental coral snakes, genus *Calliophis,* are small species from Southeast Asia that are seldom seen and apparently never bite. Most species are some shade of brown to reddish, with black bands or stripes and a white to cream band behind the head. Few species reach 3 feet in length. At the right is *Calliophis macclellandi,* while at the bottom is *C. sauteri.* Photos courtesy Dr. R. E. Kuntz.

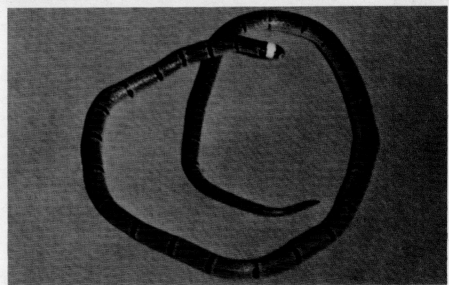

Although green mambas, *Dendroaspis angusticeps* (above and facing page), are found only in a relatively small area of southeastern central Africa, they have been extensively discussed in many books. This is probably because of their large size and reputation for being extremely dangerous, as well as the fact that it is often easier to photograph a bright green snake in a low tree or bush than a brown one under a log. The long head and very oblique scales without black edgings will usually distinguish the species.

MAMBAS

Although most elapids have relatively short and immovable fangs, the mambas are different in having longer, curved fangs that are on movable maxillary bones. The four species of *Dendroaspis* may reach 7 to 13 feet in length and are extremely agile, slender arboreal species. All mambas are found in Africa south of the Sahara, where they tend to be found mostly in brushy scrub and savannah country. They feed mostly on small mammals and birds plus the occasional lizard or frog. Eggs are laid. When excited or challenged, a mamba may spread the neck region vertically like some other tree snakes or it may spread a small hood like a cobra. It has also been reported to open the mouth widely and shake its head at an enemy.

Because of their size, the efficiency of their venom apparatus, and their occurrence in trees and bushes that puts them at head level, mambas are extremely dangerous snakes. Fortunately, they prefer to escape if possible, but they will strike rapidly if they feel cornered. Treatment is made more complicated by the fact that the various species have differing mixtures of neurotoxic and hemotoxic factors in the venom, so a single-factor antivenin may not work unless the snake can be accurately identified to species, which is not a simple task.

AUSTRALIAN ELAPIDS

The elapid snake fauna of Australia and to some extent New Guinea is diverse and abundant. Most of the genera are found only on Australia or just manage to enter New Guinea and adjacent islands (except for the sea snakes, of course, and even some of these are found only around Australia). There are about 65 species of terrestrial elapids in Australia, of which several are large enough, venomous enough, and common enough to be serious health risks.

The death adders or deaf adders, *Acantophis*, are found in Australia, New Guinea, and some adjacent Indonesian islands. These are very heavy-set snakes of moderate size (to about 3 feet), with short, slender tails and wide, triangular heads. At first glance they look very much like typical vipers, even to the point of having keeled scales (not common in elapids, which tend to be smooth and glossy snakes). The fangs are long for an elapid and the venom is very potent. The nondescript mottled and barred brownish to reddish gray color pattern makes them almost invisible on the forest floor. They are sedentary and easily stepped on by accident. Like the other Australian elapids discussed (except *Oxyuranus*), they give live birth.

The Australian copperhead, *Austrelaps superbus,* occurs only in southeastern Australia, where it is found in low-lying wet situations, often in large numbers. The color is rather plain, brownish to blackish above and cream below, but on the sides are enlarged scales often of an iridescent coppery red or yellow color. The species is seldom over about 4 feet long, but it may reach 6 feet in length. The venom is dangerous as it contains a strong hemotoxic factor as well as the usual neurotoxins.

The tiger snakes, *Notechis,* are restricted to southern Australia, where they are perhaps the most-studied venomous snakes and the ones that cause the most accidents. They are stout-bodied, dully colored snakes with rather short, thin tails. The usual length is about 4 to 5 feet. *N. ater* is solid black above with a gray belly, while *N. scutatus* is brownish with or without pale crossbars. These snakes can be active at any time of the day or night depending on the weather.

Oxyuranus scutellatus, the taipan, is the largest and most dangerous snake in Australia, reaching as it does a length of over 12 feet. It is rather plain brown above and yellowish below, with no obvious distinguishing marks. Common in northeastern Australia and New Guinea, it is found in both wet forests and dry savannahs. About 10 to 20 eggs are laid.

Several species of *Pseudechis* are found over much of Australia, but only the mulga snake or king brown snake (*P. australis*) is common. The species vary in color from brownish to iridescent black. Because of their size (to 6 to 9 feet) and daylight activity, they are dangerous species.

Although it is an elapid, the death adder, *Acanthophis antarcticus,* looks very much like a viper at first glance, especially since the head is very broad and arrowhead-shaped. The fangs are long and recurved, making it an extremely effective biter that is greatly feared throughout its rather broad range from Australia to extreme eastern Indonesia. Photo by Ken Lucas, Steinhart Aquarium.

Large tiger snakes, *Notechis scutatus,* are probably the most dangerous snakes in the heavily populated areas of southeastern Australia and cause many accidental bites each year. Photo by H. Hansen, Aquarium Berlin.

There are only two genera of American coral snakes, and both have species in the United States. *Micrurus fulvius* (right) is not uncommon in many parts of the southeastern U.S. but seldom causes accidental bites. This is one of the most colorful—if not *the* most colorful—North American snakes. The Arizona coral snake, *Micruroides euryxanthus* (below), is found in Arizona, New Mexico, and Mexico but is not commonly collected. The pattern varies considerably, but the light bands are usually white, not yellow as in *Micrurus fulvius*. Photo at right by Ken Lucas, Steinhart Aquarium; that below by Dr. Sherman A. Minton.

CORAL SNAKES

Although the name "coral snake" is given in both the New World and the Old World to many different elapids that are brightly banded in red and black, here we will restrict it to the New World species of the subfamily Micrurinae, genera *Micrurus* and *Micruroides*. With few exceptions these are slender, glossy snakes under about 3 feet in length. The head is relatively small, rounded, and not distinct from the body. Most coral snakes are marked with complete or nearly complete rings of bright red, yellow or white, and black in constant patterns that are usually specific. Although one ring color may be missing in a few species (there are about 50 species of *Micrurus* and one species of *Micruroides*), when all three colors are present they are typically arranged on the body in the sequence red-yellow-black, while nonvenomous species with similar patterns seldom have the red and yellow touching; thus the saying, with many variations, "Red touch yellow, bad for a fellow; Red touch black, good for Jack."

Coral snakes are found from the southern Atlantic Seaboard of the United States west to Arizona and south through tropical America. Most species are found in tropical forests, but a few occur in near-desert conditions. Most are nocturnal prowlers feeding on other snakes and on lizards. All are strongly venomous, but the small mouth and fangs effectively prevent them from biting humans

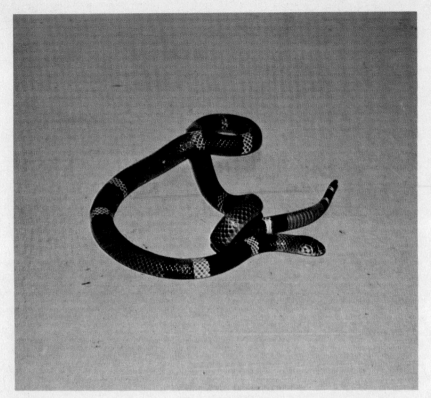

Coral snakes are usually extremely smooth and glossy animals, and many can dive into the substrate with amazing speed. The tail is usually bluntish and often is marked with two light and one dark bands (at least) that supposedly bear a mimetic relation to the head pattern. Photo of the South American *Micrurus frontalis* by Luis Mario Lozzia.

under normal circumstances. Additionally, they must hold on and chew to inoculate their venom. Almost all recorded bites have resulted from accidents in handling coral snakes, which are often very gentle and do not offer to bite unless extremely provoked. Even when they bite and chew, they seldom release much venom. However, when envenomation occurs the bite is most often fatal, as the venom is a strong and rapidly active neurotoxin. Because of their bright colors and docile temperament, they are often picked up by children, who are more easily bitten and more subject to death from even small amounts of venom than are adults. In Brazil, where at least 15 species of coral snakes occur, there were only nine fatalities in a 28-year period.

SEA SNAKES

The 50 to 60 species of sea snakes (Hydrophidinae) are something of a paradox among venomous snakes. Laboratory tests show that the venom is extremely active, probably one of the most active recorded among all the snakes. In many parts of their tropical Indo-Pacific range sea snakes are extremely abundant, occurring in groups of thousands or millions of animals. There is no doubt that a bite from even the smallest sea snake (most species are 2 to 4 feet long, with some species exceeding 6 feet in length) could be fatal. Yet sea snakes of all species are handled with impunity by fishermen throughout their range, being grabbed carelessly from fishing nets to be either killed or

Most coral snakes look much alike and at first glance are hard to separate. However, specialists can see differences in the arrangement and width of the bands over the body as well as their number. There commonly also are differences in scale counts. Shown is *Micrurus corallinus* from South America; photo by Dr. Marcos Freiberg.

One would think that the bright colors of coral snakes would make them very conspicuous, but in nature the bright pattern disappears in shadows, especially if the snake is moving. Photo of *Micrurus frontalis* by Dr. Marcos Freiberg.

In the milk snakes and most other coral snake mimics, the banding stops at the side of the body and almost never continues across the belly. In most coral snakes, however, the banding is continuous or nearly so across the entire body. Photo of *Micrurus lemniscatus* by Dr. Marcos Freiberg.

Left: *Laticauda semifasciata,* a primitive sea snake. Photo courtesy of Dr. R. E. Kuntz.

Below: *Acalyptophis peroni,* a rather small-headed species with a distinctive barred pattern. Photo by Walter Deas.

Above: Close-up of the head of an unidentified sea snake showing the dorsally placed nostrils that are closed by a valve when under water. Photo by Walter Deas. **Below:** Sea snakes are often seen by divers, especially at night. They are often very active hunters of eels and other elongated fishes, and some are not at all shy. Photo by Dr. Dwayne Reed.

thrown back into the sea. They are even common food items in some Oriental countries.

Two factors allow this cavalier treatment to occur with so few accidents—sea snakes are very hard to anger and almost never bite, and when they do bite they often inject little or no venom. When a bite occurs the symptoms are much like those of cobras and kraits. There is little or no pain or other symptoms for the first few minutes or hours after the bite. In addition to normal neurotoxic reactions, the venom also attacks muscle tissue—causing limb paralysis or pain—and destroys the kidneys. Antivenin is required for treatment of a full bite, with a recovery period of weeks in most cases and lingering effects sometimes for years.

Because of much local superstition concerning sea snakes and their bites, it is very difficult to obtain factual statistics on the number of fatalities among fishermen. Deaths from sea snake bites are often taken as a matter of course or as of religious significance and are thus not discussed freely with outsiders. In Australia, bites are uncommon but do occur, with occasional fatalities in untreated victims.

Morphologically the sea snakes are in many ways the most unusual of the snakes. Except for the relatively little-specialized *Laticauda*, all sea snakes are strongly modified for a marine existence. They mate and give birth in the sea, have the tail greatly flattened vertically and usable for propulsion, have dorsally located nostrils that can be sealed, and have seals around the mouth. Their venom is extremely toxic to fishes, especially eels. The strap-like ventral scutes typical of elapids and most other snakes are reduced or absent in most sea snakes as they no longer leave the water and are helpless on land.

Laticauda, however, is little more than a rather normal cobra-like snake that lives in the sea. It lays eggs on land, has the nostrils lateral in position as in most other snakes, and has fairly normal ventral scutes. In many ways it is intermediate between normal terrestrial elapids and the true sea snakes, and for this reason the sea snakes are considered to be only a subfamily of the Elapidae.

The brilliantly colored *Pelamis platurus* is perhaps the most widely distributed snake, being found in large numbers from the eastern coast of Africa to the western coast of Central America. Several other sea snakes also have large ranges, while some are known only from near small island chains. Among the most often indicted species in human envenomation are *Pelamis platurus, Enhydrina schistosa, Kerilia jerdoni, Lapemis* spp., and *Hydrophis* spp.

15: True Vipers

As mentioned earlier, the vipers (venomous snakes with fully hollow, nongrooved fangs and movable maxillary bones) are usually divided into two families or subfamilies, Viperinae and Crotalinae. Because the distribution of these two groups is largely nonoverlapping and so many types of each group are important venomous snakes, it is best to treat the groups separately. Here we will only cover selected genera and species of true Old World vipers of the subfamily Viperinae.

The true vipers are those vipers lacking a heat-sensing pit between the nostril and the eye. The pupil of the eye is usually vertical, separating vipers from most other snakes (except Crotalinae and many rear-fanged genera); the scales are usually strongly keeled or roughened; and the usual head plates are broken into small, irregular plates (*Causus* and *Atractaspis* have normal head plates). With few exceptions these are heavy-bodied snakes that look sluggish but can move and strike rapidly if they wish. Most species are terrestrial, but a few (such as *Atheris*) are arboreal. Few species exceed 2 to 3 feet in length, although the larger African *Bitis* species may exceed 6 feet. Most true vipers give live birth, but a few *Vipera,* all *Causus* and *Atractaspis,* and some *Echis* and *Pseudocerastes* lay eggs. Although most true vipers are subtropical in range, some of the Eurasian species may extend north to the Arctic Circle, making them among the most northerly-dwelling snakes.

SMALL VIPERS

The most familiar and best-studied viper is the common Eurasian viper or adder, *Vipera berus.* This stout-bodied species seldom exceeds 2 feet in length and is widespread throughout Europe (including Scandinavia) and extends into western Asia in southwestern Russia; other populations occur in Mongolia and extend into Japan, making it a very widely distributed species indeed. In color it is usually dark

The Near East viper, *Vipera xanthina,* is one of the more dangerous species of the genus, though certainly not in the class with *V. russelii.* In Israel it is the leading cause of snakebite. Photo by Dr. Otto Klee.

Facing page: The European viper, *Vipera berus,* is a stout-bodied little snake that can be amazingly common in even densely populated areas, causing numerous cases of snakebite each year even in England.

In western Europe one of the most common vipers is *Vipera aspis* (obviously called the asp viper), another small species that is usually considered to be harmless to man. Although the pattern normally resembles that of the other small European vipers, occasionally melanistic individuals (above) occur and can offer problems in making correct identifications. Photos courtesy Dr. D. Terver, Nancy Aquarium.

Vipera ursini is a small and in-conspicuous species from southern Europe and southwestern Russia. At under 20 inches in length and with a very weak venom, it is usually harmless to man. Photo courtesy Dr. D. Terver, Nancy Aquarium.

grayish with a zigzag blackish band down the middle of the back formed by partially fused pairs of oval spots; on the back of the head is a dark inverted "V" mark or arrowhead. The color varies considerably in different parts of its range.

Common adders are northern snakes and must face cold winters. In the late fall they move into burrows once occupied by small mammals and become torpid for the winter. These hibernation burrows are sometimes shared with small lizards and toads. With the coming of spring they become active again and start roaming the countryside in search of mates. Males are more numerous than females and compete violently for mating privileges. Collective mating is common, with many males trying to mate with one or two females. Their tangled bodies are the origin of the phrase "a nest of vipers." About 6 to 20 young (6 to 8 inches long) are born in late summer or early fall and are active and venomous as soon as they are delivered. At the northern edge of the range it may take two seasons for the embryos to develop because of the short feeding period allowed by the weather.

Common adders are shy snakes and do not try to bite unless molested. If at all possible they first try to escape. Because of their small size, the bite is not extremely serious, and few deaths result even from untreated bites. Since these snakes occur in heavily populated countries (even England) and are not uncommon near human habitation, the chance of bites is increased.

There are over a dozen species of *Vipera*, with most of the species found in the northern Middle East and adjacent Russia, with a few species in northern Africa and one species widely distributed in India and Southeast Asia. In addition to *V. berus*, the snub-nosed viper, *V. latastii*, is also found in Europe. It is restricted to the Iberian Peninsula and adjacent northwestern Africa, where it prefers less humid areas. It is much like *V. berus* in appearance but the middorsal band more distinctly forms a zigzag pattern with dark edges. Like the other vipers, it feeds mostly on small birds and mammals, with an occasional lizard or frog, plus scorpions and centipedes.

Several *Vipera* species have developed a turned up snout, sometimes with tall mounds of scales at the end of the nose. These species, such as *V. ammodytes*, are typical of dry areas, including deserts.

The most dangerous species of *Vipera* is the Russell's viper, *Vipera russelii*, of India and Southeast Asia. This heavy-bodied species reaches over 5 feet in length and has extremely strong venom that causes blood clotting. It is also one of the most attractive of the vipers, with a light yellowish to brownish body brightly marked with large dark brown ovals outlined in black and white. There is a chain-like series of these ovals down the middle of the back and another series of separate ovals on each side. This is a species of the hills and other drier areas, and it does not tend to be found near human habitation. Although shy, it

One of the most attractive of the true vipers is *Vipera russelii* of the Indian region and Southeast Asia. This strikingly patterned and very large species (for a viper) is one of the most dangerous species and causes many deaths each year, especially in India. This large female is "guarding" at least a half dozen similarly patterned young. Photo by H. Hansen, Aquarium Berlin.

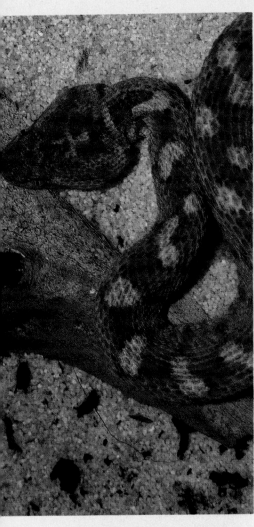

will bite repeatedly if cornered, and the venom is injected in large amounts. The death rate is high in bites that are not immediately treated with antivenin.

The deserts and plains of northern Africa, southern Asia, and India have several genera of rather small vipers such as *Eristicophis, Pseudocerastes, Cerastes,* and *Echis.* Even when they are common, however, their small size (usually only 1 or 2 feet) makes most of them relatively nondangerous to man. In India the saw-scaled viper, *Echis carinatus,* is one of the most abundant venomous snakes. It inflicts more bites than any other species, with the bites often being fatal. The venom is an anticoagulant, with death from internal bleeding resulting several days (sometimes two weeks) after the bite. The horned viper, *Cerastes cerastes,* is a small species of open sand deserts of North Africa and the Middle East and is most famed for the projecting scales over the eyes and its "sidewinding" method of locomotion by throwing the body into broad curves; both these traits are paralleled by the sidewinder of western North America, a rattlesnake.

The African bush vipers, *Atheris,* are more slender than the other vipers and are arboreal. The eight species of *Atheris* tend to be solid colors, usually brownish or greenish, are about 2 feet long, and have the scales on the head and body pointed and separated at their tips so they stand up like the scales of a pinecone. They are found in the low branches of trees and bushes in western and central Africa. Their tails are prehensile and used to help in the hunt for small tree frogs, lizards, and birds in the branches. Because they are relatively shy and docile snakes with a weak venom, they are not considered much of a threat to humans.

Although the saw-scaled vipers (*Echis*) seldom exceed 30 inches, they are among the most deadly of snakes. Both species are typical of drier regions, with *E. coloratus* (above) being found from Egypt to the Arabian Peninsula and *E. carinatus* (right) from Morocco to Ceylon. Photos by John T. Kellnhauser.

The various species of *Bitis* are usually called puff adders or just African adders and are all snakes to be avoided. Even the smaller species can inflict very severe bites, and the larger species have enough venom to kill up to five humans in a single dose. These heavy-bodied snakes are usually colored to match their surroundings and tend to be dull brownish, as is this *Bitis arietans* (shown eating a rat), though the largest species may have considerable areas of bright blue.

Facing page: The horned vipers, *Cerastes*, are true desert snakes ranging from the Sahara Desert of northern Africa to drier parts of the Middle East. The species seldom exceed 2 feet in length, are not aggressive, and are seldom a serious danger to humans. Their side-winding gait and "horns" over the eyes (often absent) are paralleled by the sidewinder rattlesnake of the southwestern United States.

BITIS

The dozen or so species of *Bitis* include some of the most feared of venomous snakes—the puff adder, rhinoceros viper, and Gaboon viper. Although many of the species of the genus are small snakes of desert and mountain regions, these three species are more typical of savannahs and forests in central and southern Africa and reach sizes in excess of 4 feet, with the Gaboon viper reported to reach 6 feet in length. They are extremely stocky snakes with gigantic heads on relatively narrow and weak necks, the large venom glands greatly swollen and making the head into a distinct arrowhead. In a large Gaboon viper the fangs can reach at least 1½ inches in length, perhaps more, and are not readily stopped even by thin leather.

The puff adder, *Bitis arietans,* is found over much of western, central, and southern Africa and is often common. When annoyed it tends to inflate its body and hiss loudly before striking, thus the common name. Like the other species considered, it is a very sedentary snake, tending to rapidly bury itself in the floor debris of the savannah or forest, the ridges on the scales holding particles of sand next to the body to increase the camouflage effect. This means that it is very likely to be stepped on by both man and domestic animals, as it depends on its camouflage ability rather than trying to escape when disturbed.

Bitis nasicornis, the rhinoceros viper, has two or three pairs of long, upright horns on the snout; it is found throughout much of central and western Africa in humid forests. The Gaboon viper, *Bitis gabonica,* ranges from the Sudan to Zululand west to Guinea, thus covering most of non-desert Africa and being found in both savannahs and wet rain forests; it may or may not have a pair of nose horns, depending on the subspecies. Both species have a colorful pattern of browns, white, and even blue arranged in diamonds and rectangles to form a complex pattern that is almost impossible to distinguish from the leafy background on which the snakes are found. *B. nasicornis* is often called the riverjack as it tends to inhabit much more humid environments than the Gaboon viper. *Bitis gabonica* reaches a length of over 6 feet, while *B. nasicornis* is smaller, not exceeding 4 feet; the latter species has a dark arrowhead-shaped mark on top of the head (the Gaboon viper lacks this mark). Both species are very sedentary and hard to disturb, so bites are few. However, because of their size and very toxic venom, fatalities often result from their bites. Even when not fatal, a bite causes tremendous tissue damage and often permanent impairment.

NIGHT VIPERS AND MOLE VIPERS

The African and Middle Eastern snakes of the genera *Causus* (night vipers) and *Atractaspis* (mole vipers) are

The two largest *Bitis* species are often confused. The rhinoceros viper, *Bitis nasicornis,* and the Gaboon viper, *B. gabonica,* have extremely similar body patterns, and both may have long or short horn-like scales on the snout. In *B. nasicornis* (shown here) the top of the head is occupied by a broad, distinctly arrowhead-shaped dark brown mark that is diagnostic. Photo above by James K. Langhammer; that at left by Ken Lucas, Steinhart Aquarium.

160

In the Gaboon viper, *B. gabonica*, the top of the head is almost uniform pale grayish to brownish, with only a narrow dark median line that may widen slightly at the tip. Photo at right by Ken Lucas, Steinhart Aquarium; that below by James K. Langhammer.

unlike other vipers in several ways, including the presence of normal plates on top of the head and round pupils. Most of the 15 or so species of mole vipers are African, but a couple of species are found in the Middle East. Most species are about 2 feet or less in length, with some reaching 3 feet. The fangs are extremely long in this genus and are not used in the usual way. The snake jabs and jerks back with the fangs (often used singly) after rotating one or both sides of the lower jaw inward to expose the teeth; this genus often gives the appearance of having the fangs projecting outside the mouth when the jaws are closed. Large specimens can cause serious bites, and *A. microlepidota* is considered to be a serious threat in the Sudan.

Night vipers are also small snakes seldom exceeding 2 feet in length. They are mostly nocturnal and nonaggressive, with a weak venom. All four species of *Causus* are restricted to tropical and southern Africa. Although bites usually occur at night when the snakes may be stepped on accidentally, they are not considered dangerous.

16: Pit Vipers

The pit vipers, comprising about 140 to 150 species in the genera *Agkistrodon* (New and Old World), *Bothrops* (New World), *Crotalus* (New World), *Lachesis* (New World), *Sistrurus* (New World), and *Trimeresurus* (Old World), are vipers with a thermosensory pit on each side of the head between the eyes and the nostril and extending into the maxillary bone. Although once assigned to full family rank, the current tendency is to express their close relationship with the true vipers by placing them as a subfamily, Crotalinae, of the Viperidae. Of the six genera, *Sistrurus* is often combined with *Crotalus,* differing only in head scalation; *Bothrops* and *Trimeresurus* are almost indistinguishable except on geographic grounds and very minor and inconsistent details of the head structure; and *Agkistrodon* has been split in the last few years into several closely related genera that will be ignored here.

For many years scientists wondered about the function of the pit between the eye and nostril and developed various theories about it, one of the more plausible being that it could detect low-frequency sound waves. Research in the 1930's, however, proved that the pit (called the loreal pit in some technical literature) is actually a sensitive mechanism for temperature perception, a kind of thermal radar through which the snakes can detect the immediate proximity (a foot or less) of warm-blooded prey—usually rodents and birds—even in total darkness. The pit not only serves to detect temperature differences of much less than one degree Fahrenheit, but it also allows the snake to aim effectively.

Like the other vipers, pit vipers have vertical pupils that go well with their usually nocturnal habits (few species are active during full daylight). Also, most species have the plates under the tail single and not divided as in the majority of colubrid snakes. Actually, the presence of the pit identifies a pit viper immediately as no other group of snakes has such an organ. (Although some boas may have pits on

In this green bamboo pit viper, possibly *Trimeresurus erythrurus*—found from India to Thailand, the labial pit is very conspicuous. Although seldom so obvious in most pit vipers, the pit is always present and in about the same position between the nostril and the eye.

Facing page: The eyelash viper, *Bothrops schlegelii,* occurs in many patterns, varying from solid green to solid yellow, with banded-reticulated patterns of reddish on gray-green as shown here fairly common. Photo by John T. Kellnhauser.

The common copperhead, *Agkistrodon contortrix*, occurs in the form of several distinct subspecies, two of which are shown here. At the top of the coil is a broad-banded copperhead, *A. c. laticinctus*, found from Kansas to central Texas. The snake at the bottom of the pile is the southern copperhead, *A. c. contortrix*, from the Mississippi Valley and southeastern U.S. Photo by G. Marcuse.

the scales of the lips, they are not between the eye and nostril.) Most vipers are heavy-bodied snakes with large, rather triangular heads. Although the majority are distinctly sedentary and terrestrial species, *Bothrops* and *Trimeresurus* both have several arboreal species with greenish coloration and prehensile tails, and some species of *Agkistrodon* are distinctly aquatic. With few exceptions (*Agkistrodon, Sistrurus*), the plates on top of the head are broken into smaller irregular scales except for plates over the eyes and on the snout. With one or two exceptions, the scales are at least weakly keeled, and usually the keeling is strong and even knob-like.

AGKISTRODON

Next to the rattlesnakes, the North American copperhead, *Agkistrodon contortrix,* is probably the most widely known North American venomous snake. It is one of about a dozen species of the genus; three *Agkistrodon* are found in North America and the other ten or so species are found in Asia. All are somewhat similar snakes with nine head plates, usually a banded pattern on the body, and typically a broad black or dark brown band running from the eye back along the head and bordered above and below by white. Several of the species are partially aquatic or show a preference for damp lowlands, while others may be found in rocky or hilly areas. In North America all three species are livebearers; most Asian species are also livebearers, with a few laying eggs.

The copperhead is widely distributed from the Great Plains through most of eastern North America to New England. There are several subspecies that vary con-

siderably in color, but most are a pale coppery tan to pinkish tan above with darker brown bands across the back, the bands usually constricted at the middle to produce hourglass-like markings. Few copperheads exceed 3 feet in length. These snakes are often abundant in hilly meadows, forested mountains, and bottomlands, and some of the subspecies inhabit near-desert conditions. Copperheads are able to maintain large numbers even near heavy human populations, which makes them the major source of snakebite in the eastern U.S. Fortunately their venom is mild, their fangs short, and their temperament usually docile. Fatalities are almost never recorded. Even if untreated, recovery from a bite is almost certain, though there may be considerable tissue damage near the bite.

The water moccasin or cottonmouth, *Agkistrodon piscivorus,* is an aquatic, very heavy-bodied species of the southern United States from central Texas to southeastern Virginia. In the South it is a common and greatly feared snake that is often confused with the myriad species of harmless brownish watersnakes (*Nerodia*) that are found in the same swamps, rivers, and lakes. Average specimens are about 3 to 4 feet in length, but some individuals may exceed 6 feet in length, making them very dangerous. When disturbed they stand their ground and open the mouth widely to show the bright white lining, the origin of the name cottonmouth. Many specimens are undistinguished dull brownish black with largely black bellies and an indistinct white line back from the eye, but in some popula-

The two water moccasins. **Above:** *Agkistrodon bilineatus,* the cantil, is a very dark tropical species with very contrasting stripes on the sides of the head. **Below:** The common cottonmouth, *A. piscivorus,* looks somewhat like a dully colored copperhead when in its best patterns, but often it is solid muddy gray-brown. Photo above by James K. Langhammer.

tions the bands are well developed and the mature animals look much like rather dull copperheads. Although deaths from even untreated bites are rare, the venom causes severe tissue damage and scarring near the bite, and recovery may take weeks.

The cantil or Mexican water moccasin, *Agkistrodon bilineatus,* is very similar in many ways to its more northern cousin but has the side of the head almost totally dark brown except for two distinct white lines above and below the eye that converge on the side of the snout. Few specimens reach 4 feet in length. It is a common species in aquatic habitats from Mexico to Nicaragua. No species of *Agkistrodon* are known to occur in southern Central America or South America.

Two Asian species of *Agkistrodon* are of major medical importance because of either their size (*A. acutus*) or wide distribution (*A. halys*). In general appearance *A. acutus* looks much like a copperhead with a distinctly upturned snout with a short upright process on it; for this reason it is called the sharp-nosed pit viper, but because of its toxicity it is also known as the 100-pacer from the widespread native belief that the victim can only run 100 paces before dying from its bite. This 4- to 5-foot-long snake is common in forested mountain areas of China, Taiwan, and Southeast Asia, where it is a very important source of human envenomations and often of human fatalities. Unlike most other species of the genus, it lays eggs.

The smaller mamushi, *A. halys,* is one of the more widely distributed species of the genus, being found as various subspecies from Japan and Korea over much of mainland China

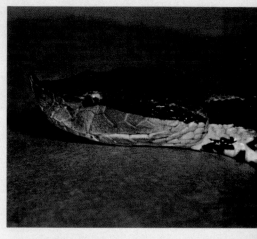

These two views of *Agkistrodon acutus* from Taiwan clearly show why it is often called the sharp-nosed pit viper. Where common, it causes many accidents and is often fatal. The pattern of dark hourglasses on a paler background seems to be an exceptionally constant one in the genus *Agkistrodon* as it is found in many species that are widely separated geographically. The same is true of the dark band back from the eye, a feature also found in many other pit vipers and vipers in general. Photos courtesy Dr. R. E. Kuntz.

Although it is one of the largest (if not *the* largest) pit vipers, the bushmaster seldom causes serious accidents. It is usually uncommon throughout its range and is avoided by natives if at all possible, so there is seldom any human-snake contact. Almost any bite by *Lachesis muta,* however, is likely to be fatal. Photo by Harald Schultz.

and adjacent Russia. It is one of the few common poisonous snakes in Japan, where it causes as many as 3,000 bites per year. Fortunately it seldom exceeds 2 feet in length and has a mild venom that almost never causes fatalities.

Two other small Asian *Agkistrodon* species are the Malayan pit viper, *A. rhodostoma,* which has smooth scales, and the hump-nosed pit viper, *A. hypnale.* Both species are typically found in plantation areas, where they cause many bites but few fatalities among the workers. *A. rhodostoma* is found over much of Southeast Asia and reaches about 2 to 3 feet in length, while *A. hypnale* is found in India and Sri Lanka and seldom exceeds 18 inches.

BUSHMASTER

Among American pit vipers the bushmaster, *Lachesis muta,* is unique in laying eggs; all other American pit vipers give live birth. It is also the largest American pit viper and probably the largest viper, reaching and probably exceeding a length of 12 feet, although 7 to 9 feet is much more common. In coloration it is pale tan or even pinkish tan with dark brownish diamonds along the middle of the back. The head is covered mostly by small irregular scales, and the tip of the tail has a long cluster of elongated scales that look somewhat like a slender pinecone. The fangs may easily exceed an inch in length, and the venom is very toxic; fortunately, the species is rare and seldom encountered. Its bite, however, is likely to be fatal. This species inhabits rain forests and lowlands from Nicaragua through most of northern South America south to Paraguay.

Of the large, long-tailed species of *Bothrops*, one of the most dangerous is the barba amarilla, *B. atrox*, of tropical America. Often reaching 6 feet in length and with very potent venom, it causes many fatalities each year. Photo by Dr. Marcos Freiberg.

BOTHROPS AND TRIMERESURUS

The many species of these two genera (about 60 *Bothrops* and 35 to 40 *Trimeresurus*) are very similar in all regards and are very difficult to identify at either the generic or specific level. *Bothrops* are all New World species and all give live birth. *Trimeresurus* are all Old World and mostly give live birth, but a few species (*T. flavoviridis* and *T. monticola,* for instance) lay eggs. In both groups the pupil is vertical and there are mostly small irregular scales on top of the head. Three general types of body forms are also shared by both genera: large, long-tailed species often with distinctive color patterns; small, very heavy-bodied, short-tailed species with indistinct brownish patterns; and small, relatively slender, prehensile-tailed greenish species that are found in trees and bushes. The large species of both genera usually have long fangs and potent venom and can cause human fatalities, but the smaller species seldom cause death. In many areas the green arboreal species are greatly feared although their bites are seldom fatal.

There are obviously so many species in these genera that only a few can be mentioned. Of the Central and South American *Bothrops* or lanceheads, the most famous and justly feared species is the barba amarilla, *Bothrops atrox* (formerly confused with the true fer-de-lance, *B. lanceolatus,* of Martinique). This 5- to 6-foot-long snake (occasional specimens reach 8 feet in length) is widely

An unidentified green palm viper from tropical America, *Bothrops* sp. It is often almost impossible to identify these featureless green snakes. Photo by G. Marcuse.

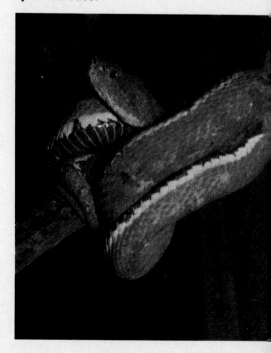

distributed from southern Mexico to Brazil and is often common in forests, in plantations, and along watercourses. It is relatively aggressive, has large fangs, and has a very toxic venom. It is probably responsible for more deaths by snakebite than any other South American species.

Generally similar in appearance and habits to *B. atrox* are a host of other species, including *B. alternatus, B. jararaca, B. neuwiedi* (with numerous subspecies), and *B. jararacussu.* These terrestrial species are usually about 3 to 4 feet long and can cause human deaths. *Bothrops jararacussu* is especially toxic and is a major cause of human deaths. The very unusual *Bothrops ammodytoides* is the southernmost viper, living only in Patagonia south to Tierra del Fuego; it has a hump of erect scales on its snout much like that of several true vipers.

Of the heavy-bodied small species, the most familiar is *Bothrops nummifer,* the jumping pit viper of Central America. This 20-inch species is common in lowland areas and is greatly feared, although its venom is not especially strong and fatalities are few. The body is so stout it is hard to believe it can find the energy to strike, but it is actually capable of rapidly striking at a distance greater than its body length, even appearing to leave the ground, thus the common name. The scales are very heavily keeled, further obscuring the pattern of indistinct brownish diamonds on a brownish background. Most of the heavy-bodied small lanceheads are Central American and northwestern South American in range.

Above: The simple, dull pattern of *Bothrops nummifer* is typical of most of the small terrestrial pit vipers of Central and South America (in fact, these specimens have better and clearer patterns than most). Photo by John T. Kellnhauser.

Right: The large *Bothrops alternatus* has a rather distinctive pattern of lateral blotches, but several related species have similar patterns. All are to be avoided, however. Photo by Dr. Marcos Freiberg.

Three relatively large South American *Bothrops* species. **Top:** *Bothrops jararaca;* photo by Dr. Marcos Freiberg. **Middle:** *Bothrops neuwiedi;* photo by Federico Achaval. **Bottom:** *Bothrops jararacussu;* photo by Dr. Marcos Freiberg.

Like most small Central American palm vipers, *Bothrops nigroviridis* is usually considered to be almost harmless because of its small size and short fangs. However, even the most "harmless" pit viper can inflict a painful bite that will require medical treatment. Photo by Dr. Sherman A. Minton.

This bright yellow specimen of *Bothrops schlegelii* clearly shows the erect, pointed scales over the eyes that earned it the name eyelash viper. Photo by John T. Kellnhauser.

Although there are many small species of *Bothrops* that have become adapted to a bush-dwelling existence and are known as palm vipers, most are obscure species with limited ranges and are seldom seen. *Bothrops schlegelii*, the eyelash viper, however, has often been imported and its green subspecies and color phases are often exhibited. This 18-inch, rather slender species ranges from Mexico to Ecuador and is often common both in virgin rain forests and in cultivated plantations. The name eyelash viper comes from the presence of a row of small pointed scales above the eye. Although most specimens seen are bright green, the species actually varies in color from tan to yellowish. Its bite is painful but usually not deadly. The rain forests of Venezuela and Colombia have an abundance of species of palm vipers of various types that bear a general resemblance to *B. schlegelii*.

Many species of *Trimeresurus* are found from India east through Southeast Asia, and several are important in human snakebite. Of the larger ground-dwelling species, *Trimeresurus mucrosquamatus* is one of the most familiar and important. It is one of several very similar species called habus from China and off-lying areas; other habus are *T. flavoviridis* of Okinawa, *T. elegans* of the Ryukyus, and *T. okinavensis* of Okinawa. Most species of habu are about 2 to 3 feet long, although *T. mucrosquamatus* may reach 5 feet. They are usually active, nocturnal prowlers that may venture into houses; they are often nervous and will bite with little provocation. The larger species are quite dangerous, with *T. mucrosquamatus* having a 7% mortality rate in Taiwan.

There is a cluster of species of *Trimeresurus* adopted to a tree-dwelling existence, with a slender body, prehensile tail, and often greenish color. These species, including especially *T. gramineus* of the Indian and Chinese area and *T. stejnegeri* of China and Taiwan, are often common in forests and bamboo thickets and are often called bamboo vipers. Few exceed 30 inches in length. *Trimeresurus stejnegeri* is unusual among snakes in that the color pattern is sexually dimorphic, a situation that once led to the creation of many synonyms. Both sexes are solid bright green above and yellow below, with a thin whitish line separating the dorsal color from the belly color. In females this line is plain, but in males there is a thinner red line just above it, a pattern sometimes very obvious in living specimens.

The extremely wide head of this Oriental habu-type pit viper, *Trimeresurus tokarensis,* is caused in large part by the great size of the venom glands plus the relatively slender neck. The systematics of the habus (and almost all *Trimeresurus* species, for that matter) are confused and species identifications are usually uncertain. Photo by John T. Kellnhauser.

Right: The himehabu, *Trimeresurus okinavensis,* a small, stoutly built, and very unaggressive species from Okinawa. Photo by John T. Kellnhauser. **Center:** *Trimeresurus mucrosquamatus,* the Taiwan habu, is a large and quite dangerous species that often causes fatalities. Photo courtesy Dr. R. E. Kuntz. **Bottom:** This female *Trimeresurus stejnegeri* lacks the thin red line above the yellowish line that would indicate a male. Photo courtesy Dr. R. E. Kuntz.

The black-tailed rattlesnake, *Crotalus molossus,* is often black at both ends. Not only is the tail velvety black, but there is often a dark brownish band across the snout as well. It is found from the Edwards Plateau of Texas to Arizona and south into Mexico. Photo by James K. Langhammer.

The southeastern subspecies of the pigmy rattlesnake, *Sistrurus miliarius barbouri,* is sometimes attractively colored but is usually dusky grayish. Photo by F. J. Dodd, Jr.

RATTLESNAKES

Rattlesnakes are instantly recognizable by the jointed, hollow rattle at the tip of the tail. Although most pit vipers and many other snakes have the habit of rapidly vibrating the tip of the tail when annoyed, which can result in quite a loud sound if done in dry leaves, the rattlesnakes of the genera *Sistrurus* and *Crotalus* are the only ones to develop the rattle. In newborn rattlers only a button, the small terminal segment, is present, but new rattles are added every time the skin is shed, often four or five times a year (for this reason the number of rattles tells you nothing about how old the snake is). The segments are held together loosely by flanges. In a few species the flanges are absent or nearly so, thus only the button is present even in adults as new segments cannot be added. The two genera are similar in all respects except head scales: *Sistrurus* has nine regular large plates on top of the head, while *Crotalus* usually has plates only over the eyes and on the snout and irregular small scales between these plates.

The pigmy rattlesnakes, three species of *Sistrurus,* are found from the southeastern United States and the Midwest south to the central plateau of Mexico. They are

usually grayish to brownish snakes with a row of large spots or blotches down the middle of the back and stripes on the side of the head. *Sistrurus miliarius* of the southeastern U.S. and *S. ravus* of the Mexican plateau seldom exceed 2 feet in length and are too small to be considered seriously dangerous, although the bite of even a newborn specimen is extremely painful and can cause severe swelling. Only the massasauga, *S. catenatus,* found from the Great Lakes south through the Great Plains and western states to northern Mexico, is large enough to inflict serious bites. It occasionally exceeds 3 feet in length, and there have been a few human fatalities from its bite. Many herpetologists, incidentally, consider *Sistrurus* to be a synonym of *Crotalus.*

The true rattlesnakes (about 30 species and many subspecies) vary in size from several dwarf species under 2 feet in length to the large diamondbacks and their relatives that may exceed 7 feet in length. Most rattlesnakes have potent venom even in small species, with some species having venom that causes severe tissue damage although seldom fatal. The bite of any rattlesnake is not to be disregarded and should be given prompt medical attention. The cut-and-suck method of first aid is still recommended by many doctors for treating the bites of North American pit vipers and seems to work better than for non-American venomous species.

The timber rattlesnake of the eastern United States, *Crotalus horridus,* reaches about 4 to 6 feet in length. It is a heavy-bodied species with dark brown to blackish cross-

Above: The massasauga, *Sistrurus catenatus,* is the only pigmy rattlesnake large enough to cause human fatalities, but these are extremely rare. It is widely distributed and often locally common in parts of its range, so bites are not uncommon. Photo by James K. Langhammer.

Right: The western pigmy rattlesnake, *Sistrurus miliarius streckeri,* has a reduced pattern compared to the more easterly subspecies. It can be extremely abundant in some areas and has a reputation for dwelling in blackberry brambles. Bites on children may be serious. Photo by F. J. Dodd, Jr.

bands on a pale tan to golden or even black background; there is no light line from the eye to the corner of the mouth, and the entire tail is black. Once widespread and common in many different habitats from New England to Texas, it has been exterminated from much of its range and is seldom found near human habitation. A bite from a large adult can easily kill a person.

The various species of diamondbacks all have much the same pattern of dark distinct diamonds, often outlined with white, on a paler brown background. There are usually two white stripes from the eye to the corner of the mouth outlining a broader dark brown or black oblique band, and the tail is often banded in dark and light. These are large species with long fangs and are capable of injecting large quantities of venom in each bite, so all are very dangerous snakes. The largest of the species is the eastern diamondback rattlesnake, *Crotalus adamanteus,* found from the Carolinas south through much of Florida and west to barely enter Louisiana. It may exceed 8 feet in length, although specimens over 6 feet are unusual today. The western diamondback, *C. atrox,* is found from Arkansas to California and into Mexico; it is usually about 5 feet long but may reach 7 feet. The red diamondback, *C. ruber,* is smaller than the first two and seldom exceeds 4 feet; it is found from Baja California to southern California. The eastern and western diamondbacks are the most deadly snakes in

The three large eastern rattlesnakes. **Above:** *Crotalus horridus,* the timber rattler. **Below left:** *Crotalus adamanteus,* the eastern diamondback; photo by F. J. Dodd, Jr. **Below right:** *Crotalus atrox,* the western diamondback; photo by F. J. Dodd, Jr. All three species are to be considered very dangerous and should be avoided if at all possible.

Three smaller western rattlesnakes. **Top:** *Crotalus scutulatus,* the Mohave rattlesnake, the technically most venomous of the North American rattlers, at least in the laboratory. **Center:** *Crotalus viridis nuntius,* a western or prairie rattler. **Bottom:** *Crotalus viridis viridis,* another subspecies of the western rattler. Photos by F. J. Dodd, Jr.

The only widespread rattlesnake in South America is the cascabel, *Crotalus durissus.* This large, aggressive, and very dangerous snake shows considerable variation in details of pattern and color, but the two dark lines running back from the head and the distinct white-outlined diamonds are usually present. The top photo is of *C. d. cascavella* from northeastern Brazil; the other two are of *C. d. terrificus* from southeastern Brazil. Photos by Dr. Marcos Freiberg.

The sidewinder, *Crotalus cerastes*, is a well-known rattler found in sandy deserts. Readily recognized by the erect scales ("horns") over the eyes, it is the American parallel of the African desert viper *Cerastes*. Photo by F. J. Dodd, Jr.

The rock rattler, *Crotalus lepidus*, is also found in the southwestern U.S. and Mexico, but it prefers rocky habitats in the dry mountains. The color pattern is one of the most distinctive of the U.S. rattlers. Photo by James K. Langhammer.

the United States and cause most of the yearly deaths from snakebite in this country. *C. ruber* is relatively mild and not normally found near human populations.

In Mexico the diamondback-type of rattler is *Crotalus basiliscus*, along with a few other less dangerous species. It seldom exceeds 5 feet in length and looks much like a faded western diamondback with the head pattern greatly obscured. It is found along much of the western Mexican coast.

Few rattlesnakes range south of Mexico, and only two species are found in South America proper (plus a small species on Aruba). However, the South American cascabel, *Crotalus durissus*, is certainly one of the most dangerous snakes on that continent. It is widely distributed, with one subspecies (*C. d. terrificus*) reaching south to northern Argentina. Reaching a length of about 5 to 6 feet, it usually has two broad dark brown stripes along the sides of the neck from the back of the head to the most anterior diamond-shaped dorsal marking. The tail is largely black. This is one of the species of rattlesnakes in which the various subspecies or geographical populations have venoms that differ considerably in action and potency. In Brazil the venom is extremely toxic, while in other parts of South America it is less so, though still deadly. Curiously, it does not form sufficient antibodies in horses to produce an effective antivenin, so large doses of relatively ineffective antivenin must be administered.

17: Duck-billed Platypuses and Insectivores

Mammals evolved from reptiles many million years ago, and all have the same general characteristics: young fed on milk, hair on the body, only one bone forming the lower jaw, and a four-chambered heart. All are warm-blooded, being able to keep their body temperature more-or-less constant regardless of the atmospheric temperatures. However, living mammals are readily divisible into three smaller groups that to some degree represent the probably gradual evolution from reptiles by several different mammalian stems. These are the monotremes (egg-laying mammals), marsupials (pouched mammals), and the true mammals from the primitive insectivores and bats through the elephants and grazing mammals. Venom and the ability to inoculate it are found in at least two types of mammals.

PLATYPUSES

The best-known venomous mammal is the duck-billed platypus, *Ornithorhynchus anatinus,* of eastern Australia. The platypus is an aquatic monotreme. It and its only living relatives, the spiny anteaters or echidnas of Australia and New Guinea, lay from one to three half-inch-diameter leathery-shelled eggs. These are incubated for one and a half to four weeks before hatching into naked young less than an inch long. Milk is secreted by glands on the mother's abdomen (the glands are exposed in platypuses but in a temporary pouch in echidnas) onto tufts of hair, where it is sucked by the young.

Monotremes have several reptilian characters but are true mammals, although primitive and not closely related to the other mammalian groups. Males of all three genera of monotremes are unique in having a strong spur on the inside of the ankles of the hind legs. In the echidnas the spur is simply capable of producing wounds by scratching or

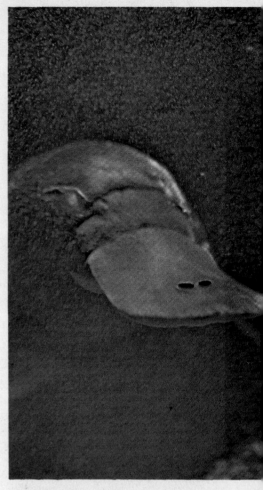

The duck-billed platypus is certainly the most unique of mammals and is familiar to every school child. Actually, however, they are very secretive and poorly studied, with many details of their lives being unknown.

Facing page: Shrews are almost exclusively nocturnal and are very secretive, so even in rural areas very few people have seen a shrew except when their cat brings one home.

slashing, but in the male duck-billed platypus the spur is hollow, has a small opening at the tip, and is connected by a duct to a venom gland.

The actual purpose of the spurs is uncertain, although they obviously have a function in defense. Curiously, the ankle spurs are present in young duck-bills of both sexes but degenerate in the females as they mature. This would indicate the spurs have a function in mating or courtship, perhaps in fights between males. Supposedly, the venom is especially strong through the Australian spring, the mating season, while less toxic during the Australian summer. In man, a wound from the spur causes intense pain and sometimes extreme swelling, with the aftereffects lingering for weeks or months before the limb is fully usable again. There is a record of a male spurring a female platypus and almost killing her. Dogs are reported to have died from being spurred while attacking platypuses.

As mentioned, the spurs in echidnas are nonvenomous, but the body spines of these animals (which look much like large hedgehogs or tailless porcupines) may possibly be mildly venomous. The spines are modified hairs and have a

Platypuses are seldom found in zoos outside Australia. In some ways their body resembles that of an otter or beaver, other aquatic mammals, but of course their head is completely different and they lay eggs, making them very strange mammals indeed.

182

hollow central shaft. Mild rashes develop quickly in humans who handle the animals, presumably from the action of the spines. Little is known of exactly what happens, however.

INSECTIVORES

Of the true mammals (excluding monotremes and the marsupials such as kangaroos and opossums), the Insectivora is the most primitive order. Though undoubtedly mammals in the normal sense of the word, they have many traits in common with some of the oldest fossils, and many types are very generalized. Of the 300 or so species of insectivores now living, the most familiar types are the moles, hedgehogs, and shrews. Some shrews are weakly venomous, and the odd solenodons of Cuba and Haiti are definitely venomous.

Solenodons, two species of the genus *Solenodon* in the family Solenodontidae, are large (to about a foot in body length plus another 7 to 12 inches of tail) and primitive insectivores that long have fascinated scientists. In general appearance they look more like very long-snouted opossums than typical shrews or moles. The tail, snout, and feet

Although at first glance the echidnas, represented here by *Tachyglossus aculeatus,* look something like porcupines, they are also egg-layers closely related to the platypus. Although the spurs are definitely not venomous in these animals, the spines of the back may cause local rashes.

The water shrew of Eurasia, *Neomys fodiens* (left and below), feeds on small frogs, fishes, and invertebrates that it kills or at least disables with its mildly venomous saliva. As in the other venomous insectivores, the venom comes from the lower jaw salivary glands.

Because of their very high metabolism, shrews must spend most of their time feeding. The mild venom probably helps give them an advantage over their prey and allows them to capture more food in a shorter time. This is a species of *Sorex*.

are almost naked, and the body pelage looks thin and scraggly. One species is found only in Cuba, the other only in Haiti; both are in danger of extinction from predation by introduced dogs and mongooses. They are secretive animals active mostly at night, and little is known about their life history.

Of major concern to us is the structure of the skull and teeth in the solenodons. The skull is almost flat, with the rostral area elongated. The teeth are heavy, and the first upper incisors are greatly enlarged and separated from the rest of the teeth by a gap. In the elongated lower jaw the teeth are also heavy, and the second lower incisors are enlarged and deeply grooved. The lower salivary glands produce a venomous saliva that drains by ducts into the grooves of the lower incisors, much like the venom apparatus in the gila monsters.

The venom is certainly used in fights between solenodons, and it is known that a solenodon bite is capable of killing another solenodon. The animals are so uncommon that there are very few instances of human envenomation, but the few cases involving handling of captive animals have caused minor pain and local swelling.

Much more common and familiar than solenodons are the abundant shrews found over much of the Northern Hemisphere. These small (usually mouse-size or smaller) insectivores are burrowers that are most active at night and are seldom seen by humans. In fact, they are perhaps most commonly seen as the prey of house cats. Cats often catch and kill shrews during their nocturnal hunts but do not eat them because of the shrews' strong scent glands. Instead, they bring them home as "playthings" to be found in the yard or on the back porch next morning. Shrews are vicious predators for their small size and have tremendous appetites for insects and insect larvae as well as worms and occasionally plant matter. In many shrews the salivary glands produce a toxic secretion that helps disable their prey, killing a large earthworm in minutes.

Shrews are relatively aggressive and will sometimes bite when cornered. In humans the bite is usually inconsequential, but there have been instances where the saliva caused minor toxic reactions, including extreme pain. There are no obvious aftereffects. Shrews often die from nervous shock when captured and handled, so bites are unlikely.

18: Poison Control Centers

Although this book has focused on venomous animals, they are only one of many possible causes of poisoning in humans. It has been estimated that just in the United States there are some 5,000,000 to 10,000,000 poisonings of all types (caused by animals, plants, drugs, household chemicals, industrial chemicals, or environmental accidents) per year, with over half of these incidents involving children under five years of age. Venomous animals are certainly just a small part of the overall number, probably so small that most physicians seldom see envenomation victims during their normal practice. Experts on venoms are few and scattered, and antivenins are usually only locally distributed in the areas of greatest threat for each type of animal venom (scorpion antivenin in Arizona or copperhead antivenins in the eastern U.S., for instance). Many types of envenomation are difficult to diagnose, such as the various spider bites.

In 1953 the first Poison Information Center (later to be known as Poison Control Centers) was established in the United States in Chicago. The idea behind this first Center and the many that followed it was to provide information on various poisons and their treatment to hospitals and physicians who could not be expected to maintain large libraries of toxicology literature or have immediate access to experienced toxicologists. Soon the system of Centers grew and began to deal directly with the public rather than just with medical establishments. Through the use of the telephone it was possible to rapidly interview a poisoning victim or the people helping him and determine just what was the cause of the poisoning incident, the symptoms, possible emergency first aid treatments, and whether in-person medical attention was necessary.

By the late 1970's almost 700 Poison Control Centers were functioning in the United States, with at least one in every state. Unfortunately all these Centers are not equal in staffing, funding, and expertise, but they are at least a start. Many are just an office and one or two staffers manning a telephone, serving mostly to direct enquiries to the appropriate hospital where questions can be more fully answered or treatment initiated. Others are regional in scope, serving entire states or parts of several states, with fully trained staffs with great experience in handling telephone enquiries about poisons, large modern libraries, and access to experts in all fields of toxicology.

Poison Control Centers are listed in the Yellow Pages of most area phone books and also in the White Pages, plus sometimes being listed under the hospitals with which they are associated. Many Centers can be reached by toll-free numbers, making access free and virtually instantaneous. Similar Centers are now found in several other countries, so in the event of a poisoning accident it is best to always check the local telephone directory to see if there is a number listed.

The person manning the telephone will ask for information about the victim and the poisoning. He should ask for the victim's name, location, telephone number from which the call is being placed (in case follow-up is necessary), background of the incident, and relation of the caller to the victim. He should also ask for the victim's age, weight, sex, and, in animal bites, the details of the accident, including such things as type of animal involved, nature and location of the bite, length of time since the bite occurred, and whether any symptoms have yet appeared. After evaluating this information he will determine whether or not any type of first aid should be administered, whether or not medical attention is necessary, and whether or not immediate transportation to a hospital is necessary; he can also determine if no treatment is necessary. If transportation is necessary, the staffer should be able to arrange for it with ambulance services or other emergency transportation if required. Poison Control Centers also may help locate and obtain antivenins from other areas for treatment of bites from unusual or imported animals.

Treatment for many animal envenomations must be started as soon as possible after the accident to be fully successful, and the use of your telephone to contact a local Poison Control Center should be the first step in the treatment. Amateur attempts at first aid without expert direction can be useless or even dangerous, so use this free access to expertise first.

Further Reading

The following selection of books and popular articles represents a minute portion of the literature on venomous animals. Many of these references have good bibliographies that will allow the interested reader to find other literature. Most should be available at your local public or college library.

Barnes, R.D. 1974. *Invertebrate Zoology*. W.B. Saunders; Philadelphia, PA.

Biery, T.L. 1976. *Venomous Arthropod Handbook*. U.S. Govt. Ptg. Off.; Washington, DC.

Bliss, S. (Editor). 1982. "Africanized bees arrive in Panama," *Smithsonian Institution Research Reports*, No. 36: 8. (Spring)

Borror, D.J. and D.M. DeLong. 1964. *An Introduction to the Study of Insects*. Holt, Rinehart & Winston; N.Y., NY.

Cargo, D.G. and L.P. Schultz. 1966. "Notes on the biology of the sea nettle, *Chrysaora quinquecirrha*, in Chesapeake Bay," *Chesapeake Sci.*, 7(2): 95-100.

Cargo, D.G. and L.P. Schultz. 1967. "Further observations on the biology of the sea nettle and jellyfishes in Chesapeake Bay," *Chesapeake Sci.*, 8(4): 209-220.

Cheng, T.C. 1973. *General Parasitology*. Academic Press; N.Y., NY.

Cogger, H.G. 1975. *Reptiles and Amphibians of Australia*. A.H. & A.W. Reed; Sydney, Australia.

Comstock, J.H. (Edited by W.J. Gertsch). 1940. *The Spider Book*. Doubleday; N.Y., NY.

Cooke, J.A.L. 1972. "Stinging hairs: a tarantula's defense," *Fauna*, No. 4: 4-8.(July/Aug.)

Edwards, J.S. 1968. "Insect assassins," *Sci. Amer.*, 202(6): 72-78. (June)

Freiberg, M. 1982. *Snakes of South America*. T.F.H. Publ.; Neptune, NJ.

Friese, U.E. 1973. "The blue-ringed octopus," *Tropical Fish Hobbyist*, 21(6): 87-95. (Feb.)

Goldstein, R.J. 1983. "Freshwater stingrays—should restrictions be imposed?" *Pet-Age*, 12(9): 17-22. (March)

Gore, R. 1976. "Those fiery Brazilian bees," *Natl. Geogr.*, 149(4): 490-501. (April)

Halstead, B.W. 1965-1970. *Poisonous and Venomous Marine Animals of the World*. Vol. 1-3. U.S. Govt. Ptg. Off.; Washington, DC.

Halstead, B.W. 1978. *Poisonous and Venomous Marine Animals of the World. Revised Edition.* Darwin Press; Princeton, NJ.

Kaston, B.J. 1970. "Comparative biology of American black widow spiders," *Trans. San Diego Soc. Nat. Hist.,* 16(3): 33-82.

Kaston, B.J. 1972. *How to Know the Spiders.* Wm. C. Brown Co.; Dubuque, IA.

Keegan, H.L. 1980. *Scorpions of Medical Importance.* Univ. Mississippi Press; Jackson, MS.

Klauber, L.M. 1956. *Rattlesnakes.* Univ. California Press; Berkeley, CA.

Levi, H.W. and L.R. Levi. 1968. *Spiders and Their Kin.* Golden Press; N.Y., NY.

Leviton, A.E. 1970. *Reptiles and Amphibians of North America.* Doubleday; N.Y., NY.

Lund, D. 1977. *All About Tarantulas.* T.F.H. Publ.; Neptune, NJ.

Mackey, V. 1983. "Tips on handling your venomous marine fishes," *Tropical Fish Hobbyist,* 32(1): 34-52. (Sept.)

Minton, S.A., Jr., et al. 1965. *Poisonous Snakes of the World.* U.S. Govt. Ptg. Off.; Washington, DC.

Pickwell, G.V. 1972. "The venomous sea snakes," *Fauna,* No. 4: 17-32. (July/Aug.)

Russell, F.E. 1971. *Poisonous Marine Animals.* T.F.H. Publ.; Neptune, NJ.

Sisson, R.F. 1974. "At home with the bulldog ant," *Natl. Geogr.,* 146(1): 62-75. (July)

Trestrail, J.H., III. 1980-1981. Information Handouts on Venomous Animals (Mimeo.). Western Michigan Poison Center; Grand Rapids, MI.

Trestrail, J.H., III. 1981. "Scorpion envenomation in Michigan—3 cases of toxic encounters with poisonous stow-aways," *Vet. and Human Toxicology,* 23(1): 8-11.

Trutnau, L. 1982. *Schlangen im Terrarium. 2. Giftschlangen.* E. Ulmer GmbH & Co.; Stuttgart, Germany.

Walker, E.P., et al. 1983. *Mammals of the World. Fourth Revised Edition.* John Hopkins Univ. Press; Baltimore, MD.

Walls, J.G. 1979. *Cone Shells: A Synopsis of the Living Conidae.* T.F.H. Publ.; Neptune, NJ.

Walls, J.G. (Editor). 1982. *Encyclopedia of Marine Invertebrates.* T.F.H. Publ.; Neptune, NJ.

Whitaker, R. 1978. *Common Indian Snakes: A Field Guide.* MacMillan India Limited; Delhi, India.

Williams, S.C. 1980. "Scorpions of Baja California, Mexico and adjacent islands," *Occas. Papers Calif. Acad. Sci.,* No. 135: 1-127.

Index
Page numbers in **bold** type indicate the presence of a photo.